Introduction to Physical Geography

Introduction to Physical Geography

Travis Gardiner

Larsen & Keller
www.larsen-keller.com

Introduction to Physical Geography
Travis Gardiner
ISBN: 978-1-64172-125-7 (Hardback)

Published by Larsen and Keller Education,
5 Penn Plaza,
19th Floor,
New York, NY 10001, USA

Cataloging-in-Publication Data

Introduction to physical geography / Travis Gardiner.
     p. cm.
Includes bibliographical references and index.
ISBN 978-1-64172-125-7
1. Physical geography. 2. Geography. I. Gardiner, Travis.
GB54.5 .I58 2019
910.02--dc23

For more information regarding Larsen and Keller Education and its products, please visit the publisher's website www.larsen-keller.com

Table of Contents

Preface   **VII**

Chapter 1 **Understanding Physical Geography**   **1**
    a. Elements of Physical Geography   2

Chapter 2 **Sub-branches of Physical Geography**   **23**
    a. Geomorphology   23
    b. Pedology   28
    c. Biogeography   30
    d. Hydrology   36
    e. Meteorology   40
    f. Climatology   48
    g. Glaciology   51
    h. Oceanography   54
    i. Coastal Geography   57
    j. Palaeogeography   58
    k. Environmental Geography   62

Chapter 3 **Tools used in Physical Geography**   **64**
    a. Remote Sensing   64
    b. Maps   70
    c. GIS   77
    d. GPS   89

Chapter 4 **The Atmosphere**   **93**
    a. Atmospheric Composition   93
    b. Atmospheric Layers   97
    c. Atmospheric Pressure   101
    d. Atmospheric Temperature   102
    e. Atmospheric Circulation   104
    f. Greenhouse Effect   106
    g. Climate Change   112

Chapter 5 **The Biosphere**   **116**
    a. Life on Earth   116
    b. Species Biodiversity   121
    c. Ecological Succession   122
    d. Ecological Pyramid   128
    e. Biogeochemical Cycle   133

Chapter 6 **The Lithosphere**   **151**
    a. Rock Cycle   151
    b. Composition of Rocks   153
    c. Plate Tectonics   166

d.	Earthquakes	182
e.	Volcanoes	183
f.	Mountain Building	184
g.	Weathering	186
h.	Erosion and Deposition	188

Chapter 7   The Hydrosphere   **193**
a.	Physical Properties of Water	193
b.	Water Cycle	202
c.	Water Distribution on Earth	209
d.	Physical Oceanography	213
e.	Ocean Current	223
f.	Ocean Surface Waves	234

Permissions

Index

Preface

Physical geography is a sub-field of geography that is concerned with the study of natural environment like the hydrosphere, geosphere, atmosphere and biosphere. Earth's landforms, glaciers and ice sheets, climate, soil and environment are areas of study in this field. Based on such specializations, physical geography can be divided into a number of sub-fields, such as hydrology, glaciology, geomorphology, climatology, meteorology, oceanography, etc. Some of the diverse tools used in physical geography include GPS, GIS, remote sensing, etc. This book is a compilation of chapters that discuss the most vital concepts in the field of physical geography. The various sub-fields of physical geography along with theoretical and technological progress that have future implications are also glanced at. For someone with an interest and eye for detail, this book covers the most significant topics in this field.

Given below is the chapter wise description of the book:

Chapter 1, Physical geography is a sub-field of geography that is concerned with the study of processes occurring in the atmosphere, geosphere, hydrosphere and the biosphere. This chapter has been carefully written to provide an introduction to physical geography and the principal elements of physical geography. **Chapter 2**, Physical geography is a vast field that can be divided into various other sub-fields, such as geomorphology, hydrology, meteorology, glaciology, oceanography, palaeogeography, etc. These diverse branches of physical geography and their principles have been thoroughly discussed in this chapter. **Chapter 3**, A number of specialized tools and technologies are used in modern physical geography to understand, describe and explain the processes and structures of the Earth. The aim of this chapter is to explore the varied tools and techniques used in physical geography, such as GIS, GPS, remote sensing and maps. **Chapter 4**, The layer of gases surrounding a planet, which are held by the force of gravity is called the atmosphere. An understanding of the Earth's atmosphere is facilitated through the study of atmospheric composition, atmospheric layers, atmospheric pressure, climate change, atmospheric temperature, atmospheric circulation and the greenhouse effect, which have been thoroughly discussed in this chapter. **Chapter 5**, The hydrosphere is the total sum of all the water found in Earth's oceans, seas, rivers, lakes, streams and ponds. The chapter closely examines the key aspects of the hydrosphere through the elucidation of the physical properties of water, water cycle, water distribution on Earth, physical oceanography, ocean current, etc. for a comprehensive understanding. **Chapter 6**, The biosphere is an ecological system comprising of living organisms and their interactions with each other and with the spheres of the Earth. The topics elaborated in this chapter will help in developing an insight into the biosphere, life on Earth, species biodiversity, ecological succession, ecological pyramid and the biogeochemical cycle. **Chapter 7**, A lithosphere is the outermost shell of a planet or a natural satellite. In case of the Earth, the lithosphere includes the crust and the upper mantle. The topics elaborated in this chapter on rock cycle, composition of rocks, plate tectonics, mountain building, weathering, etc. address the crucial aspects of the lithosphere and the various processes affecting it.

At the end, I would like to thank all those who dedicated their time and efforts for the successful completion of this book. I also wish to convey my gratitude towards my friends and family who supported me at every step.

Travis Gardiner

Understanding Physical Geography

Physical geography is a sub-field of geography that is concerned with the study of processes occurring in the atmosphere, geosphere, hydrosphere and the biosphere. This chapter has been carefully written to provide an introduction to physical geography and the principal elements of physical geography.

Physical geography is a subdiscipline of geography. It is a field of knowledge that studies natural features and phenomena on the Earth from a spatial perspective. It primarily focuses on the spatial patterns of weather and climate, soils, vegetation, animals, water in all its forms, and landforms. Physical geography also examines the interrelationships of these phenomena to human activities. This subfield of geography is academically known as the Human-Land Tradition, and has seen very keen interest and growth in the last few decades because of the acceleration of human-induced environmental degradation. Thus, physical geography's scope is much broader than the simple spatial study of nature. It also involves the investigation of how humans are influencing nature. In other words, it focuses on geography as an Earth science, making use of biology to understand global flora and fauna pattern, and mathematics and physics to understand the motion of the Earth and its relationship with other bodies in the solar system. It also includes landscape ecology and environmental geography.

Within physical geography, the Earth is often split into several spheres or environments: the atmosphere, biosphere, cryosphere, geosphere, hydrosphere, lithosphere, and pedosphere. Research in physical geography is often interdisciplinary and commonly uses the a systems approach.

Physical geography can be divided into several sub-fields, including, but not limited to the following:

- Biogeography - studying geographic patterns of species distribution and the processes that result in these patterns. The field can largely be divided into five sub-fields: island biogeography, paleobiogeography, phylogeography, phytogeography, zoogeography.

- Climatology - studying the Earth's climate. Climatology examines both the nature of micro (local) and macro (global) climates and the natural and anthropogenic influences on them. The field is sub-divided by climates of various regions, and the study of specific phenomena or time periods (e.g. paleoclimatology).

- Environmental geography - studying the spatial aspects of interactions between humans and the natural world. The sub-field bridges human and physical geography, and often requires an understanding of the dynamics of physical processes as well as the ways in which human societies conceptualize the environment. It has largely become the domain of studying environmental management or anthropogenic influences.

- Geomorphology - studying the processes by which the Earth's surface is shaped, both at the present and in the past. Geomorphology seeks to understand landform history and dynamics, and predict future changes through a combination of field observations, physical experiments, and numerical modeling.

- Hydrology - studying water quality & quantity as it moves and accumulates on the Earth.

- Landscape ecology - a sub-discipline of ecology and geography that addresses how spatial variation in the landscape affects ecological processes such as the distribution and flow of energy, materials and individuals in the environment. The main difference between biogeography and landscape ecology is that the landscape ecology is concerned the flow of energy and material and their impacts on the landscape, whereas biogeography is concerned with the spatial patterns of species and chemical cycles.

- Quaternary science - an inter-disciplinary field of study focusing on the Quaternary period, which encompasses the last 2.6 million years. The field studies the last ice age and the recent interstadial the Holocene and uses proxy evidence to reconstruct the past environments during this period to infer the climatic and environmental changes that have occurred.

Elements of Physical Geography

Elements studied in physical geography include rocks and minerals, landforms, soils, animals, plants, water, atmosphere, rivers and other water bodies, environment, climate and weather, and oceans.

Rocks

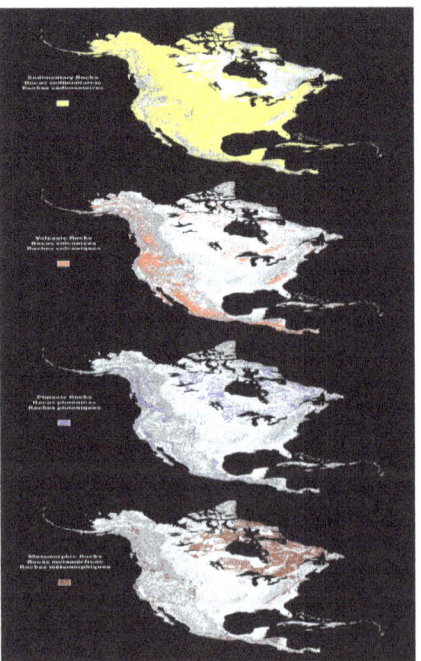

Fig; Sedimentary, volcanic, plutonic, metamorphic
rock types of North America.

A rock is a naturally occurring aggregate of minerals and mineral-like substances called mineraloids. Rocks are classified as igneous, sedimentary, and metamorphic, based on their mineral and chemical composition, the texture of the constituent particles, and the processes that formed them. The study of rocks is called petrology.

The Earth's crust (including the lithosphere) and mantle are formed of rock. The crust averages about 35 kilometers in thickness under the continents, but it averages only some 7-10 kilometers beneath the oceans. The continental crust is composed primarily of sedimentary rocks resting on crystalline "basement" formed of a great variety of metamorphic and igneous rocks, including granulite and granite. Oceanic crust is composed primarily of basalt and gabbro. Both continental and oceanic crust rest on peridotite of the Earth's mantle.

Minerals

Earth's crust is composed of many types of rocks (stones) each of which is a combination of minerals 1 or more and in geology; the condition mineral identifies any naturally occurring solid substance with a specific crystal structure and composition. The mineral's composition refers to the types and balances of elements making up the minerals. The method these elements are packed collectively determines the structure of the minerals. Many more than 3500 different minerals have been described and there are only twelve common factors (oxygen (O), silicon (Si), aluminum (Al), iron (Fe), calcium (Ca), magnesium (Mg), sodium (Na), potassium (K), titanium (Ti), hydrogen (H), manganese (Mn), phosphorus (P)) that occur in the crust of the earth. They've abundances of 0.1 percentages or more than that. Completely other naturally happening elements are found in trace amounts or very minor.

The oxygen and silicon are the most abundant crustal elements collectively comprising more than seventy percentages by weighting. It's hence not stunning that the most abundant crustal minerals are the silicates materials (for example. Mg_2SiO_4, olivine) followed by the oxides (for example. Fe_2O_3, hematite).

Types of Minerals

The two types of minerals are as follows:

1. Metallic minerals

2. Non-metallic minerals.

Metallic Minerals

The metallic minerals are those from which useful metals (for example. copper, iron) can be extracted for commercial usage. The metals that are considered geo-chemically plentiful occur at crustal abundances of 0.1 percentage or a lot (for example. titanium, manganese, iron, aluminum, magnesium).

Non-metallic Minerals

The nonmetallic minerals are useful and not for the metals they check, just for their properties as chemical combines. Since they are commonly used in manufacture, they're as well frequently referred to as minerals of industrial. They're classified according to their utilization. A few industrial minerals are practiced as sources of significant chemicals (for example. borax for borates and halite for sodium chloride).

A different important type of minarals (for example. $CaCO_3$, calcite) the sulfides (for example. PbS, galena) and the sulfates (for example. $CaSO_4$, anhydrite). Many of the large quantity minerals in the earth's crust are not of trade measure. In economically valuable minerals (nonmetallic and metallic) that provide the raw substances for industry tend to be hard and rare to determine. Hence, as considerable skill and effort is essential for finding wherever they are extracting and finding them in sufficient measures.

Formation of Minerals

Minerals are formed in various geologic environments including salt lakes, volcanoes, deep oceans, and molten rocks. Minerals can also be formed by exposing certain elements to extreme temperature and/or pressure, or by combining solutions and gases with certain concentration of specific elements. Extreme temperature and pressure alter the existing chemical composition of the elements and introduce the elements that are in the solution or gases. The suitable environment allows crystallization to occur leading to the formation of minerals.

Difference Between Minerals and Rocks

A mineral is a naturally occurring, inorganic solid with a definite atomic structure and chemical composition. A rock is an aggregation of one or several particles of minerals. One mineral can aggregates to form rocks such as quartzite while several minerals can aggregate to form rocks such as granite. Igneous rocks are formed by the crystallization of molten magma. When minerals settle and experience immense pressure, they form sedimentary rocks. When minerals are reheated and subjected to high pressure alongside pre-existing rocks, they form metamorphic rocks. Rocks do not have definite atomic structure of chemical composition.

Identifying Minerals

Some minerals are easily recognizable because of their distinctive shape and color, but some require nondestructive test to ascertain their identity. Simple tests include determining the refractive index, color, transparency, melting point, magnetism, specific gravity, radioactivity, and chemical tests. In some cases, tests involve more complex instruments such as X-ray diffractometers, energy dispersive spectrometer, and the electron microprobe.

Landforms

Land is any part of the earth's surface that is not covered by water. It is believed to cover about 30% of the earth's surface.

A landform is simply any natural geographic feature that can be found on the earth's surface, such as valleys, hills, mountains and plateau.

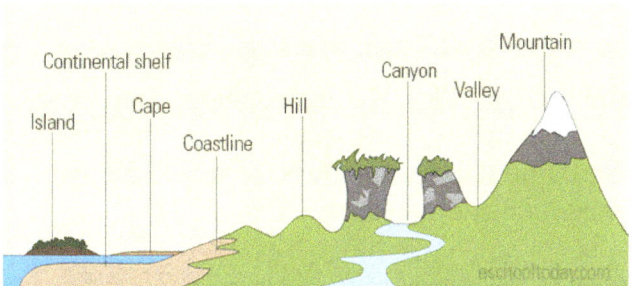

This illustration show examples of landforms

Landforms are all not the same. Some may be very high above sea level and other parts may be deep below sea level. Some of them are made of very hard material and other parts may be made of very soft material. Some landforms are covered by vegetation whiles some are void of any plant at all. Some are very large and others are small. Most important of all, landforms are constantly changing because the factors that form them are in action everyday!

Scientists who study how landforms are created, together with their interactions with other natural things are called geologists.

Formation of Landforms

Landforms are created by many factors but are all grouped into two main processes: Constructive and Destructive Processes. Landforms can be created within a short period, but some can take many decades to form. Some quick processes include landslides, volcanic eruptions, and earthquakes.

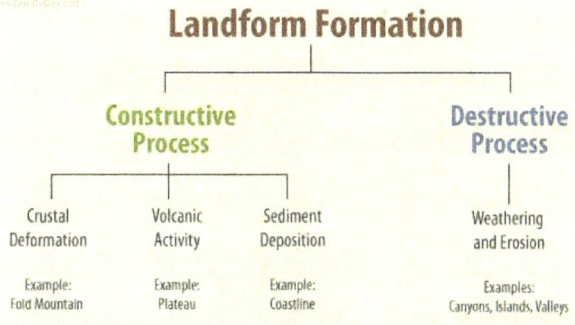

Constructive Processes

This process involves the building (addition of land material) of features either by natural action. The three main areas of constructive processes include Crustal Deformation, Volcanic Eruptions, and Sediment Deposition.

Destructive Processes

This involves the breakdown or tearing down of land surface to form new features. The process may be seen as 'carving out' parts of the land to form new features. Weathering and Erosions are the two main destructive forces that create and shape landforms.

Some Landforms to Know

Canyon

A canyon is a very steep and narrow valley. Natural elements such as water and plate tectonics can all contribute to the making of a canyon. Plate tectonics are often responsible for the formation of mountains and plateaus, and from there, water does the rest of the work. Sometimes, a river forges a path that cuts through mountains or plateaus.

A portion of the **Grand Canyon**.

Cave

A cave is a hollow space that exists deep within the Earth. Caves exist on all seven continents of the world, and even under the sea! Inside a cave, the environment stays pretty constant, with little temperature change, high humidity, and limited light sources.

The entrance to a cave running into the mountains.

Hill

A hill is a part of the landscape that's elevated above its surroundings. Hills generally aren't as steep as mountains, but they do have a high point called a summit. Hills tend to slope, and though sometimes they can be very large, they aren't as tall as a mountain.

Technically, there isn't a rule to separate hills and mountains, so sometimes it can be tricky to tell which is which. For instance, the Ozark Mountains actually consist of mountains, hills, and even plateaus!

Mountain

A mountain is one of the four major landforms. Similarly to a hill, it is an elevated portion of the Earth's surface but tends to be much larger and peaks at the top. Mountains form when tectonic plates collide at convergent boundaries, pushing both plates upwards.

Mountains protruding out from the earth's surface.

Plateau

A plateau is similar to a mountain in that it is a surface that sits elevated from sea level. However, instead of peaking at the top, it is flat. A plateau is another one of the four major landforms. Plateaus fall under different categories based on their size. A small plateau is a butte, while a medium-sized plateau is a mesa (mesa is the Spanish word for table).

Valley

When mountains form, so do valleys. A valley is a low-lying space between mountains or hills, usually shaped like the letters U or V, that is longer than it is wide. Colliding boundaries are not the only way that valleys form. Other forces of nature, such as glaciers, rivers, and streams can create valleys as well.

If a valley sits between mountains or hills, it's called a hollow. A rift valley is an especially large valley created by divergent boundaries. To diverge means to move apart or separate, so divergent boundaries are places where the tectonic plates move apart, creating a rift or gap between them.

A deep valley. It's like the opposite of a mountain.

Volcano

A volcano is a place on the Earth's surface where materials from within its core can escape. Often, a volcano takes the shape of a mountain. When you think of a mountain, you probably don't think of weak rock. However, the surface of a volcano is somewhat thin! This allows magma to break through, becoming lava once it crosses the surface.

Soils

Soil, often called the Skin of the Earth, is a mixture of decaying organic matter (humus), minerals, liquids, and many countless living organisms. Soil covering the Earth is a medium for plant growth and a means of water storage. A particular soil's texture, mineral composition, fertility, and consistency can vary from location to location.

Organisms living in the soil need water and air to thrive. These essential elements, along with the living organisms, make it possible for small animals, plants, bacteria, and fungi to live in the soil.

Soil is a key component of traditional agriculture. The two most important properties of soil, particularly in agriculture, are its ability to hold or drain water, and its nutrient composition. Additionally, all soil contains a varying mixture of three types of rock particles: clay, sand, and silt.

Different combinations of air, water, organic matter, and materials result in different soils. There are five basic soil types that most gardeners work with. These soil types are sandy, silty, clay, peaty, and saline. Each soil type varies in how it holds water, how it is managed, and how compact it is.

For example, sandy soils have a low water-holding capacity, whereas soils high in clay can have a very high water-holding capacity. The last and most ideal of soil types is loamy soil. Loamy soil is dark in color, drains well, and allows air to move freely around plant roots.

Again, as soil is an important element in the health and vitality of plants, gardeners and farmers should take care in knowing what type of soils they are using.

Plant

Plants are multicellular, mostly photosynthetic eukaryotes that also have cell walls composed of cellulose, do not have a central nervous system, are generally non-motile, and reproduce sexually, often by alternation of phases of a single generation (Alternation of generations). This kingdom includes familiar organisms such as trees, shrubs, herbs, and ferns. Over 350,000 species of plants have been estimated to exist. As of 2004, some 287,655 species had been identified, of which 258,650 are flowering plants.

In addition to plants' central ecological role—photosynthesis and carbon fixation by plants is the ultimate source of energy and organic material for nearly all ecosystems, and plants are primary producers of atmospheric oxygen—plants provide to humans vital nutritional and economic values. Indeed, the human diet is centered on plants, whether directly via grains, fruits, vegetables, legumes, and so forth, or indirectly via animals that consume or pollinate plants. Plants also provide valuable products, such as lumber, paper, and medicines. Beyond these external values, plants also touch upon the inner nature of people by providing aesthetic value and joy, such as their use in landscaping, decoration, and works of art, as well as via the smells and sights of flowers and the rich tastes of fruits.

Aristotle divided all living things between plants, which generally do not move or have sensory organs, and animals exhibiting sensory movement and motility. In Carolus Linnaeus' system, these became the Kingdoms Vegetabilia (later Plantae) and Animalia. Since then, it has become clear that the Plantae as originally defined included several unrelated groups, and the fungi and several groups of algae were removed to new kingdoms. However, these are still often considered plants in many contexts. Indeed, any attempt to match "plant" with a single taxon is doomed to fail, because plant is a vaguely defined concept unrelated to the presumed phylogenic concepts on which modern taxonomy is based.

Fern frond

Embryophytes

The most familiar plants are the multicellular land plants with specialized reproductive organs, called embryophytes. They include the vascular plants—plants with full systems of leaves, stems, and roots. They also include a few of their close relatives, often called bryophytes, of which mosses and liverworts are the most common.

All of these plants comprise eukaryotic cells with cell walls composed of cellulose, and most obtain the energy through photosynthesis, using light and carbon dioxide to synthesize food. Plants are distinguished from green algae, from which they are considered to have evolved, by having specialized reproductive organs protected by non-reproductive tissues.

Various forms of parasitism are also fairly common among plants, from the semi-parasitic mistletoe that merely takes some nutrients from its host, but still has photosynthetic leaves, to the fully parasitic broomrape and toothwort that acquire all their nutrients through connections to the roots of other plants, and so have no chlorophyll. Some plants, known as myco-heterotrophs, parasitize mycorrhizal fungi, and hence act as epiparasites on other plants.

Adiantum pedatum (a fern)

Many plants are epiphytes, meaning they grow on other plants, usually trees, without parasitizing them. Epiphytes may indirectly harm their host plant by intercepting mineral nutrients and light that the host would otherwise receive. The weight of large numbers of epiphytes may break tree limbs. Many orchids, bromeliads, ferns and mosses often grow as epiphytes. Bromeliad epiphytes accumulate water in leaf axils to form phytotelmata, complex aquatic food webs.

A few plants are carnivorous, such as the Venus Flytrap and sundew. They trap small animals and digest them to obtain mineral nutrients, especially nitrogen.

Vascular Plants

Vascular plants comprise those embryophytic plants that have specialized tissues for conducting water. Vascular plants include the seed plants—flowering plants (angiosperms), and gymnosperms—as well as non-seed (vascular) plants, such as ferns, clubmosses, and horsetails. Water transport happens in either xylem or phloem: the xylem carries water and inorganic solutes upward toward the leaves from the roots, while phloem carries organic solutes throughout the plant.

Seed plants

The spermatophytes (also known as phanerogams) comprise those plants that produce seeds. They are a subset of the embryophytes or land plants: living spermatophytes include cycads, Ginkgo, conifers, gnetae, and angiosperms.

Seed-bearing plants were traditionally divided into angiosperms, or flowering plants, and gymnosperms, which includes the gnetae, cycads, ginkgo, and conifers. Angiosperms are now thought to have evolved from a gymnosperm ancestor, which would make gymnosperms a paraphyletic group if it includes extinct taxa. Modern cladistics attempts to define taxa that are monophyletic, traceable to a common ancestor and inclusive therefore of all descendants of that common ancestor. Although not a monophyletic taxonomic unit, "gymnosperm" is still widely used to distinguish the four taxa of non-flowering, seed-bearing plants from the angiosperms.

Molecular phylogenies have conflicted with morphologically-based evidence as to whether extant gymnosperms comprise a monophyletic group. Some morphological data suggests that the Gnetophytes are the sister-group to angiosperms, but molecular phylogenies have generally shown a monophyletic gymnosperm clade that includes the Gnetophytes as sister-group to the conifers.

The fossil record contains evidence of many extinct taxa of seed plants. The so-called "seed ferns" (Pteridospermae) were one of the earliest successful groups of land plants, and forests dominated by seed ferns were prevalent in the late Paleozoic (359 - 253 mya). Glossopteris was the most prominent tree genus in the ancient southern supercontinent of Gondwana during the Permian period (299 - 253 mya). By the Triassic period (253 - 201 mya), seed ferns had declined in ecological importance, and representatives of modern gymnosperm groups were abundant and dominant through the end of the Cretaceous, when angiosperms radiated.

Modern classification classifies the seed plants as follows:

- Cycadophyta, the cycads
- Ginkgophyta, the ginkgo
- Pinophyta, the conifers
- Gnetophyta, including Gnetum, Welwitschia, Ephedra
- Magnoliophyta, the flowering plants.

Nonseed Plants

The nonseed plants are often divided into five main groups:
- Ferns (Pteridophyta or Filicophyta)
- Whisk ferns (Psilotophyta)
- Clubmosses, spikemosses, and quillworts (Lycopodiophyta)
- Horsetails (Sphenophyta or Equisetophyta)
- Adderstongues (Ophioglossophyta, but have also been grouped with the true ferns in Pteridophyta).

Pteridophyta

Pteridophyta (previously known as Filicophyta) is a vast group of 20,000 species of plants found globally, and known as ferns. Ferns can vary in complexity and size, from 2cm aquatic ferns to several-meter tree ferns of the tropics. Ferns can be either terrestrial species growing in the soil or they can be epiphytes growing on another plant. The fern life cycle differs from that of the angiosperms and gymnosperms in that its gametophyte is a free-living organism. Each frond (leaf) is capable of bearing spores (sporophyll) when conditions are right.

Psilotophyta

Psilotophyta, or Psilotales (the "whisk ferns") is a grouping of nonseed plants that sometimes is considered as an order of the Class Ophioglossopsida. This order contains only two living genera, Psilotum, a small shrubby plant of the dry tropics, and Tmesipteris, anepiphyte found in Australia, New Zealand, and New Caledonia. There has long been controversy about the relationships of the Psilotophyta, with some claiming that they are ferns (Pteridophyta), and others maintaining that they are descendants of the first vascular plants (the Psilophyta of the Devonian period). Recent evidence from DNA demonstrates a much closer relationship to the ferns, and that they are closely related to the Ophioglossales, in particular.

Psilotales lack leaves, instead having small outgrowths called enations. The enations are not considered true leaves because there is only a vascular bundle just underneath them, but not inside, as in leaves. Psilotales also do not have true roots. They are anchored by rhizoids. Absorption is aided by symbiotic fungi called mycorrhizae.

Three sporangia are united into a synangium, which is considered to be a very reduced series of branches. There is a thick tapetum to nourish the developing spores, as is typical of eusporangiate plants. The gametophyte looks like a small piece of subterranean stem, but produces antheridia and archegonia.

Ophioglossophyta

The Ophioglossophyta (lit. 'snake-tongue-leaved') are a small group of plants, the adders'-tongues and the moonworts and grape-ferns. Traditionally, they are included in the division Pteridophyta, the ferns, originally as a family and later as the order Ophioglossales. However, it is now recognized that this group is wholly distinct from the ferns and apparently from the other extant groups of plants. Thus they may be given a separate division, called the Ophioglossophyta. One scheme groups them with the horsetails and whisk ferns in the division Archeophyta.

The two principal families of ophioglossoids are the adders'-tongues, Ophioglossaceae, and the moonworts and grape-ferns, Botrychiaceae. Many workers still place the moonworts in the Ophioglossaceae, along with the distinct species Helminthostachys zeylanica. Other times, this species is given its own family Helminthostachiaceae.

All the ophioglossoids have short-lived spores formed in sporangia lacking an annulus, and borne on a stalk that splits from the leaf blade; and fleshy roots. Many species only send up one frond or leaf-blade per year. A few species send up the fertile spikes only, without any conventional leaf-blade. The gametophytes are subterranean. The spores will not germinate if exposed to sunlight, and the gametophyte can live some two decades without forming a sporophyte.

The genus Ophioglossum has the highest chromosome counts of any known plant.

Lycopodiophyta

The Division Lycopodiophyta (sometimes called Lycophyta), comprising the clubmosses, spike-mosses, and quillworts, is the oldest extant (living) vascular plant division and includes some of the most "primitive" extant species. These species reproduce by shedding spores and have macroscopic alternation of generations, although some are homosporous while others are heterosporous. They differ from all other vascular plants in having "microphylls," leaves that have only a single vascular trace (vein) rather than the much more complex megaphylls found in ferns and seed plants.

There are three main groups within the Lycopodiophyta, sometimes separated at the level of order and sometimes at the level of class. These are subdivided at the class level here:

- Class Lycopodiopsida – clubmosses and firmosses
- Class Selaginellopsida – spikemosses
- Class Isoetopsida – quillworts

The members of this division have a long evolutionary history, and fossils are abundant worldwide, especially in coal deposits. In fact, most known genera are extinct. The Silurian (444 - 417 mya) species Baragwanathia longifolia represents the earliest identifiable Lycopodiophyta, while some Cooksonia seem to be related.

The Lycopodiophyta are one of several classes of plants that expanded onto land during the Silurian and Devonian periods. They developed specialized roots to extract nutrients from the soil and developed leaves for photosynthesis and gas exchange, using a stem for transport. A waxy cuticle helped retain moisture, and stoma allowed respiration. The vulnerable meiotic gametophyte is protected from radiation by its reduced size and often by the use of subterranean mycorrhiza for its energy source instead of photosynthesis. Club-mosses are homosporous, but spike-mosses and quillworts are heterosporous. In heterospores, the female spores are larger than the male because they store food for the new generation.

Sphenophyta

The horsetails comprise 15 species of plants in the genus Equisetum. This genus is the only one in the family Equisetaceae, which in turn is the only family in the order Equisetales and the class Equisetopsida. This class is often placed as the sole member of the Division Equisetophyta (also called Arthrophyta in older works), though some recent molecular analyses place the genus within Pteridophyta, related to Marattiales. Other classes and orders of Equisetophyta are known from the fossil record, where they were important members of the world flora during the Carboniferous (359 - 299 mya) period.

The name "horsetail" arose because it was thought that the stalk resembled a horse's tail; the name Equisetum is from the Latin equus, "horse," and seta, "bristle." Other names, rarely used, include candock (applied to branching species only), and scouring-rush (applied to the unbranched or sparsely branched species). The name scouring-rush refers to its rush-like appearance and because the stems are coated with abrasive silica that led them to be used for scouring cooking pots in the past.

The genus is near-cosmopolitan, being absent only from Australasia and Antarctica. They are perennial plants, either herbaceous, dying back in winter (most temperate species) or evergreen (some tropical species, and the temperate Equisetum hyemale). They mostly grow 0.2-1.5 m (0.6 - 4.9 ft) tall, though E. telmateia can exceptionally reach 2.5 m (8.2 ft), and the tropical American species E. giganteum 5 m (16.4 ft), and E. myriochaetum 8 m (53.7 ft).

In these plants, the leaves are greatly reduced, being represented only by whorls of small, translucent scales. The stems are green and photosynthetic, also distinctive in being hollow, jointed, and ridged (with 6 - 40 ridges). There may or may not be whorls of branches at the nodes; when present, these branches are identical to the main stem except smaller.

The spores are borne in cone-like structures (strobilus, pl. strobili) at the tips of some of the stems. In many species they are unbranched, and in some (e.g., E. arvense) they are non-photosynthetic, produced early in spring separately from photosynthetic sterile stems. In some other species (e.g., E. palustre), they are very similar to sterile stems, photosynthetic, and with whorls of branches.

Horsetails are mostly homosporous, though in E. arvense, smaller spores give rise to male prothalli. The spores have four elaters that act as moisture-sensitive springs, ejecting the spores through a weak spot of the sporangia.

The horsetails were a much larger and more diverse group in the distant past before seed plants became dominant across the Earth. Some species were large trees reaching to 30 m (99.4 ft) tall. The genus Calamites (Family Calamitaceae) is abundant in coal deposits from the Carboniferous period.

Nonvascular Plants

Non-vascular plants include those land plants (embryophytes) without a vascular system. Bryophytes—the Bryophyta (mosses), the Hepaticophyta (liverworts), and the Anthocerotophyta (hornworts)—are the only nonvascular plants grouped within the Kingdom Plantae. In these groups, the primary plants are haploid, with the only diploid portion being the attached sporophyte, consisting of a stalk and sporangium. Because these plants lack water-conducting tissues, they fail to achieve the structural complexity and size of most vascular plants.

Some algae are also nonvascular, but these are no longer grouped in the plant kingdom. Recent studies have demonstrated that the algae actually consist of several unrelated groups. It turns out that common features of living in water and photosynthesis were misleading as indicators of close relationship.

Importance

The photosynthesis and carbon fixation conducted by land plants and algae are the ultimate source of energy and organic material in nearly all ecosystems. These processes radically changed the composition of the early Earth's atmosphere, which as a result is now approximately 20 percent oxygen. Animals and most other organisms are aerobic, relying on oxygen; those that do not are confined to relatively rare anaerobic, oxygen-depleted, environments.

Much of human nutrition depends on plants, whether directly or indirectly via animals that consume or pollinate plants. Much of the human diet comes in the form of cereals. Other plants or plant parts that are eaten include fruits, vegetables, legumes, herbs, and spices. Strict vegetarians rely entirely on plants (as well as some algae and fungi) for their nutrition. Many plants provide important medicines.

Some vascular plants, referred to as trees and shrubs, produce woody stems and are an important source of building material or raw material for producing paper.

Beyond these ecological, nutritional, and economic values, plants also touch upon the human inner nature through the aspect of beauty. Trees and flowering plants are used in landscaping and decoration, and are featured in works of art. The smells and sights of flowers have a valuable impact on the human mood, and the tastes of fruits bring enjoyment to people.

Distribution

Plants are found throughout the world, both on land and in water bodies. Plants are most abundant where resources (water, sunlight, adequate growth temperatures, and fertile soil) are most abundant, and accordingly, the tropics overwhelmingly contain the greatest biomass and species diversity. The mostly dry, subtropical regions contain highly specialized, dessication-tolerant species, and the plant cover is often sparse. The temperate midlatitudes once again increase in biodiversity and biomass, but for the most part do not surpass the tropics in either. Poleward of the midlatitudes, biodiversity decreases, and tundra dominates. Poleward of the Arctic Circle, vegetation growth is highly seasonal, as it remains dark for a significant portion of the year, preventing photosynthesis from occurring.

Animal

Animal is a group of multicellular eukaryotic organisms (i.e., as distinct from bacteria, their deoxyribonucleic acid, or DNA, is contained in a membrane-bound nucleus. They are thought to have evolved independently from the unicellular eukaryotes. Animals differ from members of the two other kingdoms of multicellular eukaryotes, the plants (Plantae) and the fungi (Mycota), in fundamental variations in morphology and physiology. This is largely because animals have developed muscles and hence mobility, a characteristic that has stimulated the further development of tissues and organ systems.

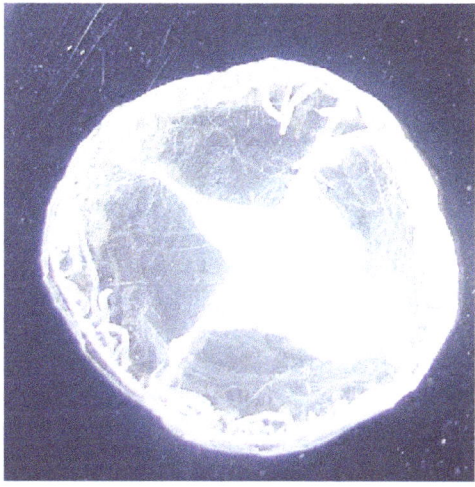

freshwater jellyfish

Animals dominate human conceptions of life on Earth not simply by their size, abundance, and sheer diversity but also by their mobility, a trait that humans share. So integral is movement to the conception of animals that sponges, which lack muscle tissues, were long considered to be plants. Only after their small movements were noticed in 1765 did the animal nature of sponges slowly come to be recognized.

In size animals are outdone on land by plants, among whose foliage they may often hide. In contrast, the photosynthetic algae, which feed the open oceans, are usually too small to be seen, but marine animals range to the size of whales. Diversity of form, in contrast to size, only impinges peripherally on human awareness of life and thus is less noticed. Nevertheless, animals represent three-quarters or more of the species on Earth, a diversity that reflects the flexibility in feeding, defense, and reproduction which mobility gives them. Animals follow virtually every known mode of living that has been described for the creatures of Earth.

Gray whale (*Eschrichtius robustus*) breaching.

Animals move in pursuit of food, mates, or refuge from predators, and this movement attracts attention and interest, particularly as it becomes apparent that the behaviour of some creatures is not so very different from human behaviour. Other than out of simple curiosity, humans study animals to learn about themselves, who are a very recent product of the evolution of animals.

A characteristic of members of the animal kingdom is the presence of muscles and the mobility they afford. Mobility is an important influence on how an organism obtains nutrients for growth and reproduction. Animals typically move, in one way or another, to feed on other living organisms, but some consume dead organic matter or even photosynthesize by housing symbiotic algae. The type of nutrition is not as decisive as the type of mobility in distinguishing animals from the other two multicellular kingdoms. Some plants and fungi prey on animals by using movements based on changing turgor pressure in key cells, as compared with the myofilament-based mobility seen in animals. Mobility requires the development of vastly more elaborate senses and internal communication than are found in plants or fungi. It also requires a different mode of growth: animals increase in size mostly by expanding all parts of the body, whereas plants and fungi mostly extend their terminal edges.

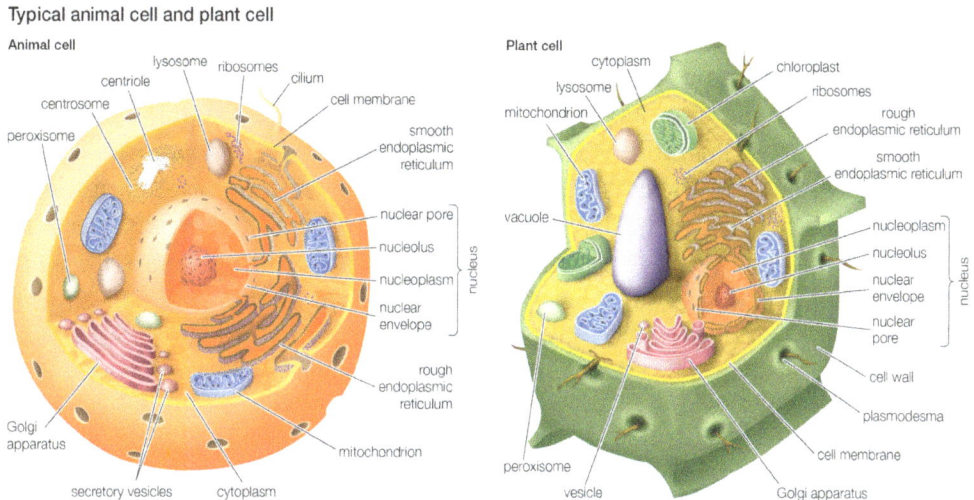

Typical animal cell and plant cell

Cytoplasm is contained within cells in the space between the cell membrane and the nuclear membrane.

All phyla of the animal kingdom, including sponges, possess collagen, a triple helix of protein that binds cells into tissues. The walled cells of plants and fungi are held together by other molecules, such as pectin. Because collagen is not found among unicellular eukaryotes, even those forming colonies, it is one of the indications that animals arose once from a common unicellular ancestor.

The muscles that distinguish animals from plants or fungi are specializations of the actin and myosin microfilaments common to all eukaryotic cells. Ancestral sponges, in fact, are in some ways not much more complex than aggregations of protozoans that feed in much the same way. Although the sensory and nervous system of animals is also made of modified cells of a type lacking in plants and fungi, the basic mechanism of communication is but a specialization of a chemical system that is found in protists, plants, and fungi. The lines that divide an evolutionary continuum are rarely sharp.

Mobility constrains an animal to maintain more or less the same shape throughout its active life. With growth, each organ system tends to increase roughly proportionately. In contrast, plants and fungi grow by extension of their outer surfaces, and thus their shape is ever changing. This basic difference in growth patterns has some interesting consequences. For example, animals can rarely sacrifice parts of their bodies to satisfy the appetites of predators (tails and limbs are occasionally exceptions), whereas plants and fungi do so almost universally.

Water

Water is the most abundant liquid on Earth. It covers more than 70% of the earth's surface. Including the clouds (which are, of course, also water), it makes our entire planet look blue and white from space.

The earth's supply of water is constantly being recycled. It is evaporated from the oceans by the sun and is given off by the forests. The vapor condenses into clouds, which rain out onto the land. The land water runs off into the lakes and rivers, which then run back to the seas, and the cycle is complete. The total amount of water on Earth, in the form of oceans, lakes, rivers, clouds, polar ice, etc. is 1.5×1018 (one-and-a-half billion) tons, occupying a total volume of 8.7 million cubic miles.

Importance

Water takes many different shapes on the earth: water vapor and clouds in the sky, waves and ice-bergs in the sea, glaciers in the mountain, and aquifers in the ground, to name but a few. Through evaporation, precipitation and runoff, water is continuously flowing from one form to another, in what is called the water cycle. Because of the importance of precipitation to agriculture, and to mankind in general, we give different names to its various forms: while rain is common in most countries, other phenomena are quite surprising when seen for the first time: hail, snow, fog or dew for example. In many African countries, snow is, for example, a very rare phenomenon. When appropriately lit, water drops in the air can refract the beautiful colours of a rainbow.

Similarly, water runoffs have played major roles in our history: rivers and irrigation supplied the water needed for agriculture. Rivers and the seas offered opportunity for travel and commerce. Through erosion, runoffs played a major part in shaping our environment providing river valleys and deltas which provide rich soil and level ground for the establishment of population centers.

Water also infiltrates the ground and goes into aquifers. This groundwater later flows back to the surface via springs, or more spectacularly via hot springs and geysers. Groundwater is also extracted artificially from wells.

Because water can contain many different substances, it can taste or smell very differently. In fact, we have developed a way to evaluate the drinkability of water: we avoid the salty seas and the putrid swamps, and we like the fresh pure water of a mountain spring.

Weather

Weather, state of the atmosphere at a particular place during a short period of time. It involves such atmospheric phenomena as temperature, humidity, precipitation (type and amount), air pressure, wind, and cloud cover. Weather differs from climate in that the latter includes the synthesis of weather conditions that have prevailed over a given area during a long time period—generally 30 years.

Weather, as most commonly defined, occurs in the troposphere, the lowest region of the atmosphere that extends from the Earth's surface to 6–8 km (4–5 miles) at the poles and to about 17 km (11 miles) at the Equator. Weather is largely confined to the troposphere since this is where almost all clouds occur and almost all precipitation develops. Phenomena occurring in higher regions of the troposphere and above, such as jet streams and upper-air waves, significantly affect sea-level atmospheric-pressure patterns—the so-called highs and lows—and thereby the weather conditions at the terrestrial surface. Geographic features, most notably mountains and large bodies of water (e.g., lakes and oceans), also affect weather. Recent research, for example, has revealed that ocean-surface temperature anomalies are a potential cause of atmospheric temperature anomalies in successive seasons and at distant locations. One manifestation of such weather-affecting interactions between the ocean and the atmosphere is what scientists call the El Niño/Southern Oscillation (ENSO). It is believed that ENSO is responsible not only for unusual weather events in the equatorial Pacific region (e.g., the exceedingly severe drought in Australia and the torrential rains in western South America in 1982–83) but also for those that periodically occur in the mid-latitudes (as, for example, the record-high summer temperatures in western Europe and unusually heavy spring rains in the central United States in 1982–83). The ENSO event of 1997–98 was associated with winter temperatures well above average in much of the United States. The ENSO phenomenon appears to influence mid-latitude weather conditions by modulating the position and intensity of the polar-front jet stream.

Generally speaking, the changeability of weather varies widely in different parts of the world. It is most pronounced in the mid-latitude belts of the westerly winds, where a usually continuous procession of traveling high- and low-pressure centres produces a constantly shifting weather pattern. In tropical regions, by contrast, weather varies little from day to day or from month to month.

Weather has a tremendous influence on human settlement patterns, food production, and personal comfort. Extremes of temperature and humidity cause discomfort and may lead to the transmission of disease; heavy rain can cause flooding, displacing people and interrupting economic activities; thunderstorms, tornadoes, hail, and sleet storms may damage or destroy crops, buildings, and transportation routes and vehicles. Storms may even kill or injure people and livestock. At sea and along adjacent coastal areas, tropical cyclones (also called hurricanes or typhoons) can cause great damage through excessive rainfall and flooding, winds, and wave action to ships, buildings, trees, crops, roads, and railways, and they may interrupt air service and communications. Heavy snowfall and icy conditions can impede transportation and increase the frequency of accidents. The long absence of rainfall, by contrast, can cause droughts and severe dust storms when winds blow over parched farmland, as with the "dustbowl" conditions of the U.S. Plains states in the 1930s.

The variability of weather phenomena has resulted in a long-standing human concern with predictions of future weather conditions and weather forecasting. In early historical times, severe weather was ascribed to annoyed or malevolent divinities. Since the mid-19th century, scientific weather forecasting has evolved, using the precise measurement of air pressure, temperature, humidity, and wind direction and speed to predict changing weather. The development of weather satellites since the 1980s has enabled meteorologists to track the movement of cyclones, anticyclones, their associated fronts, and storms worldwide. In addition, the use of radar permits the monitoring of precipitation, clouds, and tropospheric winds. To predict the weather one week or more in advance, computers combine weather models, which are based on the principles of physics, with measured weather variables, such as current temperature and wind speed. These developments have improved the accuracy of local forecasts and have led to extended and long-range forecasts, although the high variability of weather in the mid-latitudes makes longer-range forecasts less accurate. In tropical regions, by contrast, daily weather variations are small, with regularly occurring phenomena and perceptible change associated more with seasonal cycles (dry weather and monsoons). For some tropical areas, tropical cyclones themselves are one of the more influential weather variables.

Climate

The climate of a region or city is its typical or average weather. For example, the climate of Hawaii is sunny and warm. But the climate of Antarctica is freezing cold. Earth's climate is the average of all the world's regional climates.

Climate change, therefore, is a change in the typical or average weather of a region or city. This could be a change in a region's average annual rainfall, for example. Or it could be a change in a city's average temperature for a given month or season.

Climate change is also a change in Earth's overall climate. This could be a change in Earth's average temperature, for example. Or it could be a change in Earth's typical precipitation patterns.

Do you know the difference between weather and climate?
Burning coal, oil and gas to create energy releases gases into the air.

Difference Between Weather and Climate

Weather is the short-term changes we see in temperature, clouds, precipitation, humidity and wind in a region or a city. Weather can vary greatly from one day to the next, or even within the same day. In the morning the weather may be cloudy and cool. But by afternoon it may be sunny and warm.

The climate of a region or city is its weather averaged over many years. This is usually different for different seasons. For example, a region or city may tend to be warm and humid during summer. But it may tend to be cold and snowy during winter.

The climate of a city, region or the entire planet changes very slowly. These changes take place on the scale of tens, hundreds and thousands of years.

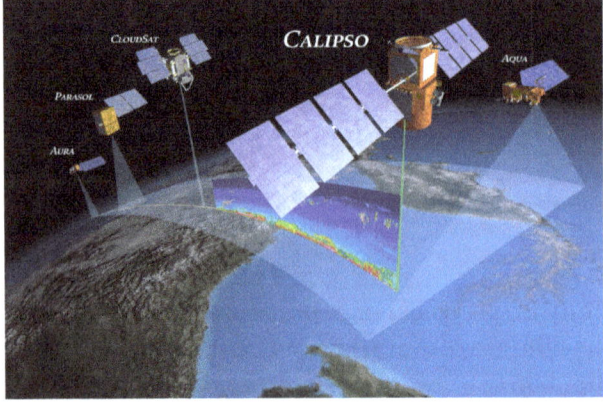

Many NASA satellites study Earth and its climate.

Changes in Earth's Climate

Earth's climate is always changing. In the past, Earth's climate has gone through warmer and cooler periods, each lasting thousands of years.

Observations show that Earth's climate has been warming. Its average temperature has risen a little more than one degree Fahrenheit during the past 100 years or so. This amount may not seem like much. But small changes in Earth's average temperature can lead to big impacts.

Causes of Earth's Climate Change

Some causes of climate change are natural. These include changes in Earth's orbit and in the amount of energy coming from the sun. Ocean changes and volcanic eruptions are also natural causes of climate change.

Most scientists think that recent warming can't be explained by nature alone. Most scientists say it's very likely that most of the warming since the mid-1900s is due to the burning of coal, oil and gas. Burning these fuels is how we produce most of the energy that we use every day. This burning adds heat-trapping gases, such as carbon dioxide, into the air. These gases are called greenhouse gases.

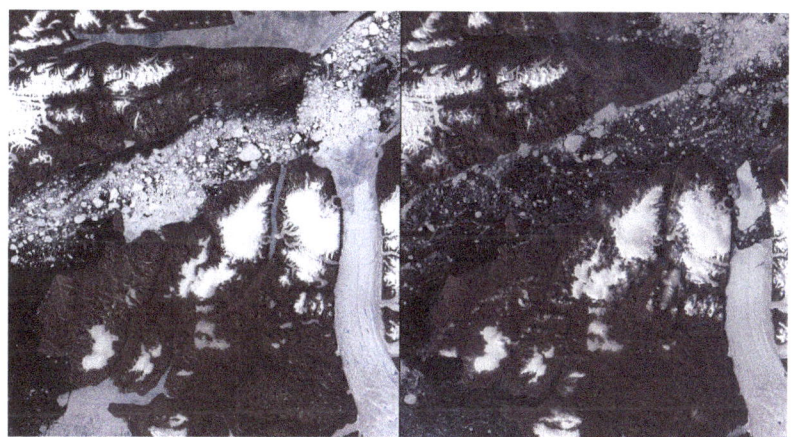

The left side of this picture is Petermann Glacier in Greenland. A huge iceberg broke off the glacier. Warmer water below the floating ice and at the sea's surface were probably caused the break.

Forecast for Earth's Climate

Scientists use climate models to predict how Earth's climate will change. Climate models are computer programs with mathematical equations. They are programmed to simulate past climate as accurately as possible. This gives scientists some confidence in a climate model's ability to predict the future.

Climate models predict that Earth's average temperature will keep rising over the next 100 years or so. There may be a year or years where Earth's average temperature is steady or even falls. But the overall trend is expected to be up.

Earth's average temperature is expected to rise even if the amount of greenhouse gases in the atmosphere decreases. But the rise would be less than if greenhouse gas amounts remain the same or increase.

Impact of Earth's Warming Climate

Some impacts already are occurring. For example, sea levels are rising, and snow and ice cover is decreasing. Rainfall patterns and growing seasons are changing.

Further sea-level rise and melting of snow and ice are likely as Earth warms. The warming climate likely will cause more floods, droughts and heat waves. The heat waves may get hotter, and hurricanes may get stronger.

References

- Physical-geography, department-specialties: unr.edu, Retrieved 12 June 2018

- What-are-minerals, science: swiftutors.com, Retrieved 22 May 2018

- What-is-a-landform, landforms: eschooltoday.com, Retrieved 18 March 2018

- Animal: britannica.com, Retrieved 28 June 2018

- What-is-water-341: deltawerken.com, Retrieved 28 June 2018

- Weather, science: britannica.com, Retrieved 20 July 2018

Sub-branches of Physical Geography

Physical geography is a vast field that can be divided into various other sub-fields, such as geomorphology, hydrology, meteorology, glaciology, oceanography, palaeogeography, etc. These diverse branches of physical geography and their principles have been thoroughly discussed in this chapter.

Geomorphology

Geomorphology is the study of landforms, their processes, form and sediments at the surface of the Earth (and sometimes on other planets). Study includes looking at landscapes to work out how the earth surface processes, such as air, water and ice, can mould the landscape. Landforms are produced by erosion or deposition, as rock and sediment is worn away by these earth-surface processes and transported and deposited to different localities. The different climatic environments produce different suites of landforms. The landforms of deserts, such as sand dunes and ergs, are a world apart from the glacial and periglacial features found in polar and sub-polar regions. Geomorphologists map the distribution of these landforms so as to understand better their occurrence.

Earth-surface processes are forming landforms today, changing the landscape, albeit often very slowly. Most geomorphic processes operate at a slow rate, but sometimes a large event, such as a landslide or flood, occurs causing rapid change to the environment, and sometimes threatening humans. So geological hazards, such as volcanic eruptions, earthquakes, tsunamis and landslides, fall within the interests of geomorphologists. Advancements in remote sensing from satellites and GIS mapping has benefited geomorphologists greatly over the past few decades, allowing them to understand global distributions.

Geomorphologists are also "landscape-detectives" working out the history of a landscape. Most environments, such as Britain and Ireland, have in the past been glaciated on numerous occasions, tens and hundreds of thousands of years ago. These glaciations have left their mark on the landscape, such as the steep-sided valleys in the Lake District and the drumlin fields of central Ireland. Geomorphologists can piece together the history of such places by studying the remaining landforms and the sediments – often the particles and the organic material, such as pollen, beetles, diatoms and macrofossils preserved in lake sediments and peat, can provide evidence on past climate change and processes.

So geomorphology is a diverse discipline. Although the basic geomorphological principles can be applied to all environments, geomorphologists tend to specialise in one or two areas, such aeolian (desert) geomorphology, glacial and periglacial geomorphology, volcanic and tectonic geomorphology, and even planetary geomorphology. Most research is multi-disciplinary, combining the knowledge and perspectives from two contrasting disciplines, combining with subjects as diverse as ecology, geology, civil engineering, hydrology and soil science.

Landforms and Processes

Two concerns are foremost within the realm of geomorphology, and these concerns reflect the stages of its history. First, in line with Davis's original conception of geomorphology as an area of science devoted to classifying and describing natural features, there is its concern with topography. The latter may be defined as the configuration of Earth's surface, including its relief (elevation and other in equalities) as well as the position of physical features.

These physical features are called landforms, examples of which include mountains, plateaus, and valleys. Geomorphology always has involved classification, and early scientists working in this subdiscipline addressed the classification of landforms. Other systems of classification, however, are not so concerned with cataloging topographical features themselves as with differentiating the processes that shaped them. This brings us to the other area of interest in geomorphology: the study of how landforms came into being.

Shaping the Earth

Among the processes that drive the shaping of landforms is plate tectonics, or the shifting of large, movable segments of lithosphere (the crust and upper layer of Earth's mantle). Plate tectonics is discussed in detail within its own essay and more briefly in other areas throughout this book, as befits its status as one of the key areas of study in the earth sciences.

Other processes also shape landforms. Included among these processes are weathering, the breakdown of rocks and minerals at or near the surface of Earth due to physical or chemical processes; erosion, the movement of soil and rock due to forces produced by water, wind, glaciers, gravity, and other influences; and mass wasting or mass movement, the transfer of earth material, by processes that include flow, slide, fall, and creep, down slopes. Also of interest are fluvial and eolian processes (those that result from water flow and wind, respectively) as well as others related to glaciers and coastal formations.

Human activity also can play a significant role in shaping Earth. This effect may be direct, as when the construction of cities, the building of dams, or the excavation of mines alters the landscape. On the other hand, it can be indirect. In the latter instance, human activity in the biosphere exerts an impact, as when the clearing of forest land or the misuse of crop land results in the formation of a dust bowl.

Real-Life Application

Subsidence

Subsidence refers to the process of subsiding (settling or descending), on the part of either an air column or the solid earth, or, in the case of solid earth, to the resulting formation or depression. Subsidence in the atmosphere is discussed briefly in the entry Convection. Subsidence that occurs in the solid earth, known as geologic subsidence, is the settling or sinking by a body of rock or sediment. (The latter can be defined as material deposited at or near Earth's surface from a number of sources, most notably preexisting rock.)

As noted earlier, many geomorphologic processes can be caused either by nature or by human beings. An example of natural subsidence takes place in the aftermath of an earthquake, during which large areas of solid earth may simply drop by several feet. Another example can be observed at the top of a volcano some time after it has erupted, when it has expelled much of its material (i.e., magma) and, as a result, has collapsed.

Natural subsidence also may result from cave formation in places where underground water has worn away limestone. If the water erodes too much limestone, the ceiling of the cave will subside, usually forming a sinkhole at the surface. The sinkhole may fill with water, making a lake; the formation of such sinkholes in many spots throughout an area (whether the sinkholes become lakes or not), is known as karst topography.

In places where the bedrock is limestone—particularly in the sedimentary basins of rivers—karst topography is likely to develop. The United States contains the most extensive karst region in the world, including the Mammoth cave system in Kentucky. Karst topography is very pronounced in the hills of southern China, and karst landscapes have been a prominent feature of Chinese art for centuries. Other extensive karst regions can be found in southern France, Central America, Turkey, Ireland, and England.

Man-Made Subsidence

Man-made subsidence often ensues from the removal of groundwater or fossil fuels, such as petroleum or coal. Groundwater removal can be perfectly safe, assuming the area experiences sufficient rainfall to replace, or recharge, the lost water. If recharging does not occur in the necessary proportions, however, the result will be the eventual collapse of the aquifer, a layer of rock that holds groundwater.

In so-called room-and-pillar coal mining, pillars, or vertical columns, of coal are left standing, while the areas around them are extracted. This method maintains the ceiling of the "room" that has been mined of its coal. After the mine is abandoned, however, the pillar eventually may experience so much stress that it breaks, leading to the collapse of the mined room. As when the ceiling of a cave collapses, the subsidence of a coal mine leaves a visible depression above ground.

Uplift

As its name implies, uplift describes a process and results opposite to those of subsidence. In uplift the surface of Earth rises, owing either to a decrease in downward force or to an increase in upward force. One of the most prominent examples of uplift is seen when plates collide, as when India careened into the southern edge of the Eurasian landmass some 55 million years ago. The result has been a string of mountain ranges, including the Himalayas, Karakoram Range, and Hindu Kush, that contain most of the world's tallest peaks.

Plates move at exceedingly slow speeds, but their mass is enormous. This means that their inertia (the tendency of a moving object to keep moving unless acted upon by an outside force) is likewise gargantuan in scale. Therefore, when plates collide, though they are moving at a rate equal to only a few inches a year, they will keep pushing into each other like two automobiles crumpling in a head-on collision. Whereas a car crash is over in a matter of seconds, however, the crumpling of continental masses takes place over hundreds of thousands of years.

When sea floor collides with sea floor, one of the plates likely will be pushed under by the other one, and, likewise, when sea floor collides with continental crust, the latter will push the sea floor under. This results in the formation of volcanic mountains, such as the Andes of South America or the Cascades of the Pacific Northwest, or volcanic islands, such as those of Japan, Indonesia, or Alaska's Aleutian chain.

Isostatic Compensation

In many other instances, collision, compression, and extension cause uplift. On the other hand, as noted, uplift may result from the removal of a weight. This occurs at the end of an ice age, when glaciers as thick as 1.9 mi. (3 km) melt, gradually removing a vast weight pressing down on the surface below.

This movement leads to what is called isostatic compensation, or isostatic rebound, as the crust pushes upward like a seat cushion rising after a person is longer sitting on it. Scandinavia is still experiencing uplift at a rate of about 0.5 in. (1 cm) per year as the after-effect of glacial melting from the last ice age. The latter ended some 10,000 years ago, but in geologic terms this is equivalent to a few minutes' time on the human scale.

Islands

Geomorphology, as noted earlier, is concerned with landforms, such as mountains and volcanoes as well as larger ones, including islands and even continents. Islands present a particularly interesting area of geomorphologic study. In general, islands have certain specific characteristics in terms of their land structure and can be analyzed from the standpoint of the geosphere, but particular islands also have unique ecosystems, requiring an interdisciplinary study that draws on botany, zoology, and other subjects.

In addition, there is something about an island that has always appealed to the human imagination, as evidenced by the many myths, legends, and stories about islands. Some examples include Homer's Odyssey, in which the hero Odysseus visits various islands in his long wanderings; Thomas More's Utopia, describing an idealized island republic; Robinson Crusoe, by Daniel Defoe, in which the eponymous hero lives for many years on an island with no companion but the trusty native Friday; Treasure Island, by Robert Louis Stevenson, in which the island is the focus of a treasure hunt; and Mark Twain's Adventures of Huckleberry Finn, depicting Jackson Island in the Mississippi River, to which Huckleberry Finn flees to escape "civilization."

One of the favorite subjects of cartoonists is that of a castaway stranded on a desert island, a mound of sand with no more than a single tree. Movies, too, have long portrayed scenarios, from the idyllic to the brutal, that take place on islands, particularly deserted ones, a notable example being Cast Away (2000). A famous line by the English poet John Donne (1572-1631) warns that "no man is an island," implying that many wish they could enjoy the independence suggested by the concept of an island. Within the Earth system, however, nothing is fully independent, and, as we shall see, this is certainly the case where islands are concerned.

The Islands of Earth

Earth has literally tens of thousands of islands. Just two archipelagos (island chains), those that make up the Philippines and Indonesia, include thousands of islands each. While there are just a few dozen notable islands on Earth, many more dot the planet's seas and oceans. The largest are these:

- Greenland (Danish, northern Atlantic): 839,999 sq. mi.(2,175,597 sq km)

- New Guinea (divided between Indonesia and Papua New Guinea, western Pacific): 316,615 sq. mi. (820,033 sq km)

- Borneo (divided between Indonesia and Malaysia, western Pacific): 286,914 sq. mi. (743,107 sq km)

- Madagascar (Malagasy Republic, western Indian Ocean): 226,657 sq. mi. (587,042 sq km)

- Baffin (Canadian, northern Atlantic): 183,810 sq. mi. (476,068 sq km)

- Sumatra (Indonesian, northeastern Indian Ocean): 182,859 sq. mi. (473,605 sq km)

The list could go on and on, but it stops at Sumatra because the next-largest island, Honshu (part of Japan), is less than half as large, at 88,925 sq. mi. (230,316 sq km). Clearly, not all islands are created equal, and though some are heavily populated or enjoy the status of independent nations (e.g., Great Britain at number eight or Cuba at number 15), they are not necessarily the largest. On the other hand, some of the largest are among the most sparsely populated.

Of the 32 largest islands in the world, more than a third are in the icy northern Atlantic and Arctic, with populations that are small or practically nonexistent. Greenland's population, for instance, was just over 59,000 in 1998, while that of Baffin Island was about 13,200. On both islands, then, each person has about 14 frozen sq. mi. (22 sq km) to himself or herself, making them among the most sparsely populated places on Earth.

Continents, Oceans and Islands

Australia, of course, is not an island but a continent, a difference that is not related directly to size. If Australia were an island, it would be by far the largest. Australia is regarded as a continent, however, because it is one of the principal landmasses of the Indo-Australian plate, which is among a handful of major continental plates on Earth. Whereas continents are more or less permanent (though they have experienced considerable rearrangement over the eons), islands come and go, seldom lasting more than 10 million years. Erosion or rising sea levels remove islands, while volcanic explosions can create new ones, as when an eruption off the coast of Iceland resulted in the formation of an island, Surtsey, in 1963.

Islands are of two types, continental and oceanic. Continental islands are part of continental shelves (the submerged, sloping ledges of continents) and may be formed in one of two ways. Rising ocean waters either cover a coastal region, leaving only the tallest mountains exposed as islands or cut off part of a peninsula, which then becomes an island. Most of Earth's significant islands are continental and are easily spotted as such, because they lie at close proximity to continental landmasses. Many other continental islands are very small, however; examples include the barrier islands that line the East Coast of the United States. Formed from mainland sand brought to the coast by rivers, these are technically not continental islands, but they more clearly fit into that category than into the grouping of oceanic islands.

Oceanic islands, of which the Hawaiian-Emperor island chain and the Aleutians off the Alaskan coast are examples, form as a result of volcanic activity on the ocean floor. In most cases, there is a region of high volcanic activity, called a hot spot, beneath the plates, which move across the hot spot. This is the situation in Hawaii, and it explains why the volcanoes on the southern islands are still active while those to the north are not: the islands themselves are moving north across the hot spot. If two plates converge and one subducts a deep trench with a parallel chain of volcanic islands may develop. Exemplified by the Aleutians, these chains are called island arcs.

Island Ecosystems

The ecosystem, or community of all living organisms, on islands can be unique owing to their separation from continents. The number of life-forms on an island is relatively small and can encompass some unusual circumstances compared with the larger ecosystems of continents. Ireland, for instance, has no native snakes, a fact "explained" by the legend that Saint Patrick drove them away. Hawaii and Iceland are also blessedly free of serpents.

Oceanic islands, of course, tend to have more unique ecosystems than do continental islands. The number of land-based animal life-forms is necessarily small, whereas the varieties of birds, flying insects, and surrounding marine life will be greater owing to those creatures' mobility across water. Vegetation is relatively varied, given the fact that winds, water currents, and birds may carry seeds.

Nonetheless, ecosystems of islands tend to be fairly delicate and can be upset by the human introduction of new predators (e.g., dogs) or new creatures to consume plant life (e.g., sheep). These changes sometimes can have disastrous effects on the overall balance of life on islands. Overgrazing may even open up the possibility of erosion, which has the potential of bringing an end to an island's life.

Pedology

Pedology is a scientific discipline concerned with all aspects of soils, including their physical and chemical properties, the role of organisms in soil production and in relation to soil character, the description and mapping of soil units, and the origin and formation of soils.

Accordingly, pedology embraces several subdisciplines, namely, soil chemistry, soil physics, and soil microbiology. Each employs a sophisticated array of methods and laboratory equipment not unlike that used in studies of the physics, chemistry, or microbiology of nonsoils systems. Sampling, description, and mapping of soils is considerably simpler, however. A soil auger is used to obtain core samples in places where no subsurface exposure can be found, and the soil units are defined, delineated, and mapped in a manner similar to procedures in stratigraphy. Such soils studies, in fact, overlap the concerns of the stratigrapher and the geologist, both of whom may treat the soils layers as strata of the Quaternary Period (from 2.6 million years ago to the present).

Soil is not only a support for vegetation, but it is also the zone beneath our feet (the pedosphere) of numerous interactions between climate (water, air, temperature), soil life (micro-organisms, plants, animals) and its residues, the mineral material of the original and added rock, and its position in the landscape. During its formation and genesis, the soil profile slowly deepens and develops characteristic layers, called 'horizons', while a steady state balance is approached.

Soil users (such as agronomists) showed initially little concern in the dynamics of soil. They saw it as medium whose chemical, physical and biological properties were useful for the services of agronomic productivity. On the other hand, pedologists and geologists did not initially focus on the agronomic applications of the soil characteristics (edaphic properties) but upon its relation to the nature and history of landscapes. Today, there's an integration of the two disciplinary approaches as part of landscape and environmental sciences.

Pedologists are now also interested in the practical applications of a good understanding of pedogenesis processes (the evolution and functioning of soils), like interpreting its environmental history and predicting consequences of changes in land use, while agronomists understand that the cultivated soil is a complex medium, often resulting from several thousands of years of evolution. They understand that the current balance is fragile and that only a thorough knowledge of its history makes it possible to ensure its sustainable use.

Concepts

- Complexity in soil genesis is more common than simplicity.

- Soils lie at the interface of Earth's atmosphere, biosphere, hydrosphere and lithosphere. Therefore, a thorough understanding of soils requires some knowledge of meteorology, climatology, ecology, biology, hydrology, geomorphology, geology and many other earth sciences and natural sciences.

- Contemporary soils carry imprints of pedogenic processes that were active in the past, although in many cases these imprints are difficult to observe or quantify. Thus, knowledge of paleoecology, palaeogeography, glacial geology and paleoclimatology is important for the recognition and understanding of soil genesis and constitute a basis for predicting the future soil changes.

- Five major, external factors of formation (climate, organisms, relief, parent material and time), and several smaller, less identifiable ones, drive pedogenic processes and create soil patterns.

- Characteristics of soils and soil landscapes, e.g., the number, sizes, shapes and arrangements of soil bodies, each of which is characterized on the basis of soil horizons, degree of internal homogeneity, slope, aspect, landscape position, age and other properties and relationships, can be observed and measured.

- Distinctive bioclimatic regimes or combinations of pedogenic processes produce distinctive soils. Thus, distinctive, observable morphological features, e.g., illuvial clay accumulation in B horizons, are produced by certain combinations of pedogenic processes operative over varying periods of time.

- Pedogenic (soil-forming) processes act to both create and destroy order (anisotropy) within soils; these processes can proceed simultaneously. The resulting soil profile reflects the balance of these processes, present and past.

- The geological Principle of Uniformitarianism applies to soils, i.e., pedogenic processes active in soils today have been operating for long periods of time, back to the time of appearance of organisms on the land surface. These processes do, however, have varying degrees of expression and intensity over space and time.

- A succession of different soils may have developed, eroded and/or regressed at any particular site, as soil genetic factors and site factors, e.g., vegetation, sedimentation, geomorphology, change.

- There are very few old soils (in a geological sense) because they can be destroyed or buried by geological events, or modified by shifts in climate by virtue of their vulnerable position at the surface of the earth. Little of the soil continuum dates back beyond the Tertiary period and most soils and land surfaces are no older than the Pleistocene Epoch. However, preserved/lithified soils (paleosols) are an almost ubiquitous feature in terrestrial (land-based) environments throughout most of geologic time. Since they record evidence of ancient climate change, they present immense utility in understanding climate evolution throughout geologic history.

- Knowledge and understanding of the genesis of a soil is important in its classification and mapping.

- Soil classification systems cannot be based entirely on perceptions of genesis, however, because genetic processes are seldom observed and because pedogenic processes change over time.

- Knowledge of soil genesis is imperative and basic to soil use and management. Human influence on, or adjustment to, the factors and processes of soil formation can be best controlled and planned using knowledge about soil genesis.

- Soils are natural clay factories (clay includes both clay mineral structures and particles less than 2 μm in diameter). Shales worldwide are, to a considerable extent, simply soil clays that have been formed in the pedosphere and eroded and deposited in the ocean basins, to become lithified at a later date.

Biogeography

Biogeography is the science which deals with geographic patterns of species distribution and the processes that result in such patterns.

One broad pattern, for example, is that the large, native mammals in Australia are all marsupials, whereas almost all large mammals on other continents are placentals (Luria et al. 1981). This is despite the fact that many of the Australian marsupials share similar ecological roles and similarities in form to various placentals.

Darwin used biogeography as one of his principal proofs of the evolutionary theory of descent with modification, that organisms have descended from common ancestors, with each species arising in a single geographic location from another species that preceded it in time.

Biogeography essentially examines the geographic distribution of species and the various geological, evolutionary, climatic, and ecological conditions that influenced this distribution. The patterns of species distribution can usually be explained through a combination of historical factors such as speciation, extinction, continental drift, glaciation (and associated variations in sea level, river routes, and so on), and river capture, in combination with the area and isolation of landmasses (geographic constraints) and available energy supplies.

Classification

Biogeography is a synthetic science, related to geography, biology, soil science, geology, climatology, ecology, and evolution.

Some fundamental factors in biogeography are:

- Evolution (change in genetic composition of a population)

- Extinction (disappearance of a species)

- Dispersal (movement of populations away from their point of origin, related to migration)

- Range and distribution

- Endemic areas

Divisions of Biogeography

Phylogeny

Phylogenetics is the study of evolutionary relatedness among various groups of organisms (e.g., species, populations). Also known as phylogenetic systematics, phylogenetics treats a species as a group of lineage-connected individuals over time. Phylogenetic taxonomy, which is an offshoot of, but not a logical consequence of, phylogenetic systematics, constitutes a means of classifying groups of organisms according to degree of evolutionary relatedness.

Phylogeny (or phylogenesis) is the origin and evolution of a set of organisms, usually a set of species. A major task of systematics is to determine the ancestral relationships among known species (both living and extinct). The most commonly used methods to infer phylogenies include parsimony, maximum likelihood, and MCMC-based Bayesian inference. Distance-based methods construct trees based on overall similarity which is often assumed to approximate phylogenetic relationships. All methods depend upon an implicit or explicit mathematical model describing the evolution of characters observed in the species included, and are usually used for molecular phylogeny where the characters are aligned nucleotide or amino acid sequences.

Gene Transfer

Organisms can generally inherit genes in two ways: by speciation (vertical gene transfer), from parent to offspring, or by horizontal or lateral gene transfer, in which genes jump between unrelated organisms, a common phenomenon in prokaryotes.

Lateral gene transfer has complicated the determination of phylogenies of organisms since inconsistencies have been reported depending on the gene chosen.

Carl Woese came up with the three domain theory of life (Eubacteria, Archaea and Eukaryotes) based on his discovery that the genes encoding ribosomal RNA are ancient and distributed over all lineages of life with little or no lateral gene transfer. Therefore rRNA are commonly recommended as molecular clocks for reconstructing phylogenies.

This has been particularly useful for the phylogeny of microorganisms, to which the species concept does not apply and which are too morphologically simple to be classified based on phenotypic traits.

Taxon Sampling

Due to the development of advanced sequencing techniques in molecular biology, it has become feasible to gather large amounts of data (DNA or amino acid sequences) to estimate phylogenies. For example, it is not rare to find studies with character matrices based on whole mitochondrial genomes. However, it has been proposed that it is more important to increase the number of taxa in the matrix than to increase the number of characters, because the more taxa, the more robust is the resulting phylogeny. This is partly due to the breaking up of long branches. It has been argued that this is an important reason to incorporate data from fossils into phylogenies where possible.

Using simulations, Zwickl and Hillis found that increasing taxon sampling in phylogenetic inference has a positive effect on the accuracy of phylogenetic analyses.

Island Biogeography

The study of island biogeography is a field within biogeography that attempts to establish and explain the factors that affect the species diversity of a particular community. In this context, the island can be any area of habitat surrounded by areas unsuitable for the species on the island; not just true islands surrounded by ocean, but also mountains surrounded by deserts, lakes surrounded by dry land, forest fragments surrounded by human-altered landscapes.

Theory of Island Biogeography

The theory of island biogeography, also known as the equilibrium theory of island biogeography (ETIB), holds that the number of species found on an island (the equilibrium number) is determined by two factors, the effect of distance from the mainland and the effect of island size. These would affect the rate of extinction on the islands and the level of immigration.

Islands closer to the mainland are more likely to receive immigrants from the mainland than those further away from the mainland. The equilibrium number of an island close to Africa is going to be larger than that of one found in the mid-Atlantic. This is the distance effect. The size effect reflects a long known relationship between island size and species diversity. On smaller islands that chance of extinction is greater than on larger ones. Thus larger islands can hold more species than smaller ones. The play between these two factors can be used to establish how many species an island can hold at equilibrium.

The theory of island biogeography was tested by Wilson and his student Daniel Simberloff in the mangroves off Florida. Small islands of mangroves were surveyed then fumigated with methyl bromide to clear their insect and arthropod communities. The islands were then monitored to study the immigration of species to the islands (the experimental equivalent of the creation of new islands). Within a year, the islands had been recolonised, and had reached equilibrium, with islands closer to the mainland having more species, as predicted.

Island Biogeography and Conservation

Within a few years of the publishing of the theory of island biogeography, its application to the field of conservation biology had been realized and was being vigorously debated in ecological circles. The realization that reserves and national parks formed islands inside human-altered landscapes (habitat fragmentation), and that these reserves could lose species as they 'relaxed towards equilibrium' (that is they would lose species as they achieved their new equilibrium number, known as ecosystem decay) caused a great deal of concern. This is particularly true when conserving larger species, which tend to have larger ranges.

A study by Newmark showed a strong correlation between the size of a protected National Park, in the United States, and the number of species of mammals. This led to the debate known as single large or several small (SLOSS), described by writer David Quammen as "ecology's own genteel version of trench warfare." This debate involved whether a single large reserve was preferable or a combination of several smaller reserves of equal area.

In the years after the publication of Wilson and Simberloff's papers on their work in mangrove swamps ecologists had found more examples of the species area-relationship, and conservation planning was taking the view that the one large reserve could hold more species that several smaller reserves, and that larger reserves should be the norm in reserve design. This view was in particular championed by Jared Diamond. This led to concern by other ecologists, including Dan Simberloff himself. Simberloff, a proponent of the theory whose research with Wilson had provided support, reversed his views and considered the theory to be an unproven over-simplification that would damage conservation efforts. In particular, he expressed concern that applying the theory to nature reserves, involving the SLOSS perspective, would pose risks since it was unproven and lacked strong empirical support. Habitat diversity was as or more important than size in determining the number of species protected.

In species diversity, island biogeography most describes allopatric speciation. Allopatric speciation is where new gene pools arise in isolated gene pools. Island biogeography is also useful in considering sympatric speciation, the idea of different species arising from one ancestral species in the same area. Interbreeding between the two differently adapted species would prevent speciation, but in some species, sympatric speciation appears to have occurred.

Phylogeography

Phylogeography is the study of the processes controlling the geographic distributions of lineages by constructing the genealogies of populations and genes. This term was introduced to describe geographically structured genetic signal within a single species. An explicit focus on a species' biogeographical past sets phylogeography apart from classical population genetics. Phylogeographical inferences are usually made by studying the reconstructed genealogical histories of individual genes (gene trees) sampled from different populations. Past events that can be inferred include population expansion, population bottlenecks, vicariance, and migration. One of the goals of phylogeographic analyses is to evaluate the relative role of history in shaping the genetic structure of populations relative to important ongoing processes. Approaches integrating genealogical and distributional information can address the relative roles of different historical forces in shaping current patterns.

Development

While the term phylogeography was first coined in 1987, it has existed as a field of study for much longer. Historical biogeography addresses how historical, geological, climatic, and ecological conditions influenced the current distribution of species. As part of historical biogeography, researchers had been evaluating the geographical and evolutionary relationships of organisms years before. Two developments during the 1960s and 1970s were particularly important in laying the groundwork for modern phylogeography; the first was the spread of cladistic thought, and the second was the development of plate tectonics theory. The resulting school of thought was vicariance biogeography, which explained the origin of new lineages through geological events like the drifting apart of continents or the formation of rivers. When a continuous population (or species) is divided by a new river or a new mountain range (i.e., a vicariance event), two populations (or species) are created. Paleogeography, geology and paleoecology are all important fields that supply information that is integrated into phylogeographic analyses.

Phylogeography takes a population genetic and phylogenetic perspective on biogeography. In the mid-1970s, population genetic analyses turned to mitochondrial markers. The advent of the polymerase chain reaction (PCR), the process where millions of copies of a DNA segment can be replicated, was crucial in the development of phylogeography. Thanks to this breakthrough, the information contained in mitochondrial DNA sequences was much more accessible. Advances in both laboratory methods that allowed easier sequencing of DNA and computational methods that make better use of the data have helped improve phylogeographic inference. The development of coalescent theory has also played an important role.

Early phylogeographic work was sometimes criticized for its narrative nature and lack of statistical rigor. Hypothesis testing was rarely done, and the explanation of genealogical patterns was essentially story telling. Recent approaches have taken a stronger statistical approach to phylogeography that was done initially. Statistical phylogeography has received an increasing amount of attention.

Example

Climate change, such as the glaciation cycles of the past 2.4 million years, has periodically restricted some species into disjunct refugia. These restricted ranges may result in population bottlenecks that reduce genetic variation. Once a reversal in climate change allows for rapid migration out of refugial areas, these species spread rapidly into newly available habitat. A number of empirical studies find genetic signatures of both animal and plant species that support this scenario of refugia and postglacial expansion. This has occurred both in the tropics, as well as temperate regions that were influenced by glaciers.

Phylogeography and Conservation

Phylogeography can help in the prioritization of areas of high value for conservation. Phylogeographic analyses have also played an important role in defining evolutionary significant units (ESU), a unit of conservation below the species level that is often defined on unique geographic distribution and mitochondrial genetic patterns.

A somewhat surprising result of a phylogenetic analysis with high conservation value was the finding that the African elephant was in fact two divergent species, the forest elephant (Loxodonta cyclotis) and the savannah elephant (Loxodonta africana). Another recent study on imperiled cave crayfish in the Appalachian Mountains of eastern North America demonstrates how phylogenetic analyses can aid in recognizing conservation priorities. Using phylogeographical approaches, the authors found that hidden within what was thought to be a single, widely distributed species, an ancient and previously undetected species was also present. Conservation decisions can now be made to ensure that both lineages received protection. Results like this are not an uncommon outcome from phylogeographic studies.

An analysis of salamanders of the genus Eurycea, also in the Appalachians, found that the current taxonomy of the group greatly underestimated species level diversity. The authors of this study also found that patterns of phylogeographic diversity were more associated with historical (rather than modern) drainage connections, indicating that major shifts in the drainage patterns of the region played an important role in the generation of diversity of these salamanders. A thorough understanding of phylogeographic structure will thus allow informed choices in prioritizing areas for conservation.

Comparative Phylogeography

The field of comparative phylogeography seeks to accomplish a variety of objectives. For example, comparisons across multiple taxa can clarify the histories of biogeographical regions. For instance, phylogeographic analyses of terrestrial vertebrates on the Baja California peninsula and marine fish on both the Pacific and gulf sides of the peninsula display genetic signatures that suggest a vicariance event effected multiple taxa during the Pleistocene or Pliocene.

Phylogeography also gives an important historical perspective on community composition. History is relevant to regional and local diversity in two ways. One, the size and makeup of the regional species pool results from the balance of speciation and extinction. Two, at a local level community composition is influenced by the interaction between local extinction of species' populations and recolonization. A comparative phylogenetic approach in the Australian Wet Tropics indicates that regional patterns of species distribution and diversity are largely determined by local extinctions and subsequent recolonizations corresponding to climatic cycles.

Human Phylogeography

Phylogeography has also proven to be useful in understanding the origin and dispersal patterns of our own species, Homo sapiens. Based primarily on observations of skeletal remains of ancient human remains and estimations of their age, anthropologists proposed two competing hypotheses about human origins. The first hypothesis is referred to as the Out-of-Africa with replacement model, which contends that the last expansion out of Africa around 100,000 years ago resulted in the modern humans displacing all previous Homo spp. populations in Eurasia that were the result of an earlier wave of emigration out of Africa. The multiregional scenario claims that individuals from the recent expansion out of Africa intermingled genetically with those human populations of more ancient African emigrations.

A phylogeographic study that uncovered a Mitochondrial Eve that lived in Africa 150,000 years ago provided early support for the Out-of-Africa model. While this study had its shortcomings, it received significant attention both within scientific circles and a wider audience. A more thorough phylogeographic analysis that used ten different genes instead of a single mitochondrial marker indicates that at least two major expansions out of Africa after the initial range extension of Homo erectus played an important role shaping the modern human gene pool and that recurrent genetic exchange is pervasive. These findings strongly demonstrated Africa's central role in the evolution of modern humans, but also indicated that the multiregional model had some validity.

Phylogeography of Viruses

Viruses are informative in understanding the dynamics of evolutionary change due to their rapid mutation rate and fast generation time. Phylogeography is a useful tool in understanding the origins and distributions of different viral strains. A phylogeographic approach has been taken for many diseases that threaten human health, including dengue fever, rabies, influenza and HIV. Similarly, a phylogeographic approach will likely play a key role in understanding the vectors and spread of avian influenza (HPAI H5N1), demonstrating the relevance of phylogeography to the general public.

Hydrology

Hydrology is the study of water on Earth. Hydrologists look at the properties of water, the ways in which it is distributed, and the effects of water on the Earth's surface, with a goal of understanding the complex and interconnected systems which dictate life on Earth. This field does not generally include the world's oceans; rather, they are studied by oceanographers, although a hydrologist may sometimes be asked to analyze water samples from the ocean.

As you might imagine, hydrology has a number of applications. Hydrologists work on flood control programs, irrigation schemes, and hydroelectric power generation plans. They also research water for both domestic and industrial supply, and they often make up part of a team on projects ranging from construction of skyscrapers to pollution remediation. Many hydrologists choose a unique area of focus in their work, becoming specialists on issues like groundwater contamination and river flow.

A major focus in the study of hydrology is the hydrosphere, the series of interconnected water systems on Earth. Activity in the hydrosphere causes water to constantly circulate in a process called the hydrologic cycle. The hydrologic cycle moves water through the ground, along the surface of the Earth, and in the sky, retooling water molecules for new purposes on a daily basis. The water you drink, for example, might have been drunk by another human or animal at some point in its history, and it may have sat for centuries locked deep in the ground or it might have landed in your reservoir with a batch of rain last week. The study of this cycle and the things which interrupt it is a major cornerstone of the field of hydrology, as you might well imagine.

Hydrologic Cycle and Transport

The central theme of hydrology is that water moves throughout the Earth by different pathways and at different rates. The most striking image of this is in the evaporation of water from the ocean, to form clouds. These clouds drift over the land and produce rain. The rainwater flows into lakes, rivers, or aquifers. The water in lakes, rivers, and aquifers then either evaporates into the atmosphere or eventually flows back to the ocean, completing a cycle.

Moreover, water movement is a significant means by which other material, such as soil or pollutants, is transported from place to place. The initial input in receiving waters may arise from a point source discharge or a line source or area source, such as surface runoff. Since the 1960s, rather complex mathematical models have been developed, facilitated by the availability of high speed computers. The most common pollutant classes analyzed are nutrients, pesticides, total dissolved solids, and sediment.

Branches of Hydrology

- Chemical hydrology is the study of the chemical characteristics of water. It examines how water is affected as it comes into contact with different materials on and below the Earth's surface. This field includes studies on the mechanisms by which salts are transported by such processes as erosion, runoff, evaporation, and precipitation.

- Ecohydrology is the study of ecological processes in the hydrologic cycle. As these processes occur in the soil and plant foliage, ecohydrologists study how the hydrologic system affects plant physiology, soil moisture, and plant diversity and spatial orientation in various regions over a period of time. Ecohydrology has four main components: infiltration of precipitation into the soil, evapotranspiration, leakage of water into deeper portions of the soil not accessible to the plant, and runoff from the ground surface.

- Hydrogeology (or geohydrology) is the study of the distribution and movement of water in aquifers and shallow porous media—that is, the porous layers of rock, sand, silt, and gravel below the Earth's surface. Hydrogeology examines the rate of diffusion of water through these media as the water moves down its energy gradient. The flow of water in the shallow subsurface is also pertinent to the fields of soil science, agriculture, and civil engineering. The flow of water and other fluids (hydrocarbons and geothermal fluids) in deeper formations is relevant to the fields of geology, geophysics, and petroleum geology.

- Hydroinformatics is the adaptation of information technology to hydrology and water resources applications. Its purpose is to facilitate decision-making for many critical applications. Hydrological data are collected, stored, processed, and analyzed using modeling techniques and simulations, based on the knowledge of particular systems. Three common types of hydrological data collected are: the rate of flow of major rivers and streams, precipitation, and water height in wells.

- Hydrometeorology is the study of the transfer of water and energy between land and water body surfaces and the lower atmosphere. Hydrometeorology incorporates meteorology to solve hydrological problems. These problems include forcasting flood or drought, or determining water resources and the safety of dams. Hydrometeorologists try to determine, through empirical data or theory, how the dynamics of water in the atmosphere affect the greatest levels of precipitation reaching the ground. The domain of hydrometeorology in the physical sciences is not very clearly defined, as it involves cloud physics, climatology, weather forecasting, and hydrology, to name a few.

- Hydromorphology is the study of the physical characteristics of bodies of water on the Earth's surface, including river basins, channels, streams, and lakes. Water quality, levels of pollution, and biological components needed for ecological system maintenance are a few areas assessed when classifying water systems. Hydromorphology studies the dynamics of groundwater flow into channels, lakes, and streams. It measures flow patterns and geometry as well as routing flows to avoid flooding or drought.

- Isotope hydrology is the study of the isotopic signatures of water. This subfield of hydrology utilizes isotopic dating to determine the origin and age of water throughout its movement within the hydrologic cycle. Isotopic dating involves measuring the levels of deviation in the isotopes of oxygen and hydrogen in water. Researchers are able to determine groundwater dated as far back as the Ice Age by using these techniques. Isotope hydrology deals with water usage policy, mapping aquifers, conservation of water resources, and maintaining pollution levels. One way isotopic hydrology is applied today is in the mitigation of arsenic levels in the drinking water of Bangladesh.

- Surface-water hydrology is the study of bodies of water on or near the Earth's surface. Rivers, dams, lakes, and reservoirs are all part of this area of study, which further includes the systems used in recreational activity and transportation. Surface hydrology addresses issues pertaining to eroding soils and streams due to surface flow. Flooding, nutrient run-off, and pollutants are a few of the effects addressed, as well as the destruction of civil constructions such as dams. Methods of hydraulic and hydrologic design regulation are also undertaken in this field of study, as researchers simulate the long and short-term effects of anthropogenically manipulated surface water forms.

Hydrologic Measurements

The movement of water through the Earth can be measured in a number of ways. This information is important for both assessing water resources and understanding the processes involved in the hydrologic cycle. The following is a list of devices used by hydrologists and what they are used to measure:

- Disdrometer - precipitation characteristics

- Symon's evaporation pan - evaporation

- Infiltrometer - infiltration

- Piezometer - groundwater pressure and, by inference, groundwater depth

- Radar - cloud properties

- Rain gauge - rain and snowfall
- Satellite - topographic patterns of surface water
- Sling psychrometer - humidity
- Stream gauge - stream flow
- Tensiometer - soil moisture
- Time domain reflectometer - soil moisture

Hydrologic Prediction

Observations of hydrologic processes are used to make predictions of future water movement and quantity.

Statistical Hydrology

By analyzing the statistical properties of hydrologic records, such as rainfall or river flow, hydrologists can estimate future hydrologic phenomena. This approach, however, assumes that the characteristics of the processes remain unchanged.

These estimates are important for engineers and economists so that they can perform proper risk analysis for future decisions in infrastructure and to determine the yield reliability characteristics of water supply systems. Statistical information is utilized to formulate operating rules for large dams that are part of systems set up to meet agricultural, industrial, and residential demands.

Hydrologic Modeling

Hydrologic models are simplified, conceptual representations of a part of the hydrologic cycle. They are primarily used for hydrologic prediction and for understanding hydrologic processes. Two major types of hydrologic models can be distinguished:

- Models based on data: These models use mathematical and statistical concepts to link a certain input (such as rainfall) to the model output (such as runoff). These models are known as stochastic hydrology models.
- Models based on process descriptions: These models try to represent the physical processes observed in the real world. Typically, they contain representations of surface runoff, subsurface flow, evapotranspiration, and channel flow, but they can be far more complicated. These models are known as deterministic hydrology models.

Applications of Hydrology

- Mitigating and predicting the risk of flood, landslide, and drought.
- Designing irrigation schemes and managing agricultural productivity.
- Providing drinking water.
- Designing dams for water supply or hydroelectric power generation.

- Designing bridges.

- Designing sewers and urban drainage system.

- Analyzing the impacts of antecedent moisture on sanitary sewer systems.

- Predicting geomorphological changes, such as erosion or sedimentation.

- Assessing the impacts of natural and anthropogenic environmental change on water resources.

- Assessing contaminant transport risk and establishing environmental policy guidelines.

Meteorology

Meteorology is the study of the atmosphere, atmospheric phenomena, and atmospheric effects on our weather. The atmosphere is the gaseous layer of the physical environment that surrounds a planet. Earth's atmosphere is roughly 100 to 125 kilometers (65-75 miles) thick. Gravity keeps the atmosphere from expanding much farther.

Meteorology is a subdiscipline of the atmospheric sciences, a term that covers all studies of the atmosphere. A subdiscipline is a specialized field of study within a broader subject or discipline. Climatology and aeronomy are also subdisciplines of the atmospheric sciences. Climatology focuses on how atmospheric changes define and alter the world's climates. Aeronomy is the study of the upper parts of the atmosphere, where unique chemical and physical processes occur. Meteorology focuses on the lower parts of the atmosphere, primarily the troposphere, where most weather takes place.

Meteorologists use scientific principles to observe, explain, and forecast our weather. They often focus on atmospheric research or operational weather forecasting. Research meteorologists cover several subdisciplines of meteorology to include: climate modeling, remote sensing, air quality, atmospheric physics, and climate change. They also research the relationship between the atmosphere and Earth's climates, oceans, and biological life.

Scales of Meteorology

Weather occurs at different scales of space and time. The four meteorological scales are: microscale, mesoscale, synoptic scale, and global scale. Meteorologists often focus on a specific scale in their work.

Microscale Meteorology

Microscale meteorology focuses on phenomena that range in size from a few centimeters to a few kilometers, and that have short life spans (less than a day). These phenomena affect very small geographic areas, and the temperatures and terrains of those areas.

Microscale meteorologists often study the processes that occur between soil, vegetation, and surface water near ground level. They measure the transfer of heat, gas, and liquid between these surfaces. Microscale meteorology often involves the study of chemistry.

Tracking air pollutants is an example of microscale meteorology. MIRAGE-Mexico is a collaboration between meteorologists in the United States and Mexico. The program studies the chemical and physical transformations of gases and aerosols in the pollution surrounding Mexico City. MIRAGE-Mexico uses observations from ground stations, aircraft, and satellites to track pollutants.

Mesoscale Meteorology

Mesoscale phenomena range in size from a few kilometers to roughly 1,000 kilometers (620 miles). Two important phenomena are mesoscale convective complexes (MCC) and mesoscale convective systems (MCS). Both are caused by convection, an important meteorological principle.

Convection is a process of circulation. Warmer, less-dense fluid rises, and colder, denser fluid sinks. The fluid that most meteorologists study is air. (Any substance that flows is considered a fluid.) Convection results in a transfer of energy, heat, and moisture—the basic building blocks of weather.

In both an MCC and MCS, a large area of air and moisture is warmed during the middle of the day—when the sun angle is at its highest. As this warm air mass rises into the colder atmosphere, it condenses into clouds, turning water vapor into precipitation.

An MCC is a single system of clouds that can reach the size of the state of Ohio and produce heavy rainfall and flooding. An MCS is a smaller cluster of thunderstorms that lasts for several hours. Both react to unique transfers of energy, heat, and moisture caused by convection.

The Deep Convective Clouds and Chemistry (DC3) field campaign is a program that will study storms and thunderclouds in Colorado, Alabama, and Oklahoma. This project will consider how convection influences the formation and movement of storms, including the development of lightning. It will also study their impact on aircraft and flight patterns. The DC3 program will use data gathered from research aircraft able to fly over the tops of storms.

Synoptic Scale Meteorology

Synoptic-scale phenomena cover an area of several hundred or even thousands of kilometers. High- and low-pressure systems seen on local weather forecasts, are synoptic in scale. Pressure, much like convection, is an important meteorological principle that is at the root of large-scale weather systems as diverse as hurricanes and bitter cold outbreaks.

Low-pressure systems occur where the atmospheric pressure at the surface of the Earth is less than its surrounding environment. Wind and moisture from areas with higher pressure seek low-pressure systems. This movement, in conjunction with the Coriolis force and friction, causes the system to rotate counter-clockwise in the Northern Hemisphere and clockwise in the Southern Hemisphere, creating a cyclone. Cyclones have a tendency for upward vertical motion. This allows moist air from the surrounding area to rise, expand and condense into water vapor, forming clouds. This movement of moisture and air causes the majority of our weather events.

Hurricanes are a result of low-pressure systems (cyclones) developing over tropical waters in the

Western Hemisphere. The system sucks up massive amounts of warm moisture from the sea, causing convection to take place, which in turn causes wind speeds to increase and pressure to fall. When these winds reach speeds over 119 kilometers per hour (74 miles per hour), the cyclone is classified as a hurricane.

Hurricanes can be one of the most devastating natural disasters in the Western Hemisphere. The National Hurricane Center, in Miami, Florida, regularly issues forecasts and reports on all tropical weather systems. During hurricane season, hurricane specialists issue forecasts and warnings for every tropical storm in the western tropical Atlantic and eastern tropical Pacific. Businesses and government officials from the United States, the Caribbean, Central America, and South America rely on forecasts from the National Hurricane Center.

High-pressure systems occur where the atmospheric pressure at the surface of the Earth is greater than its surrounding environment. This pressure has a tendency for downward vertical motion, allowing for dry air and clear skies.

Extremely cold temperatures are a result of high-pressure systems that develop over the Arctic and move over the Northern Hemisphere. Arctic air is very cold because it develops over ice and snow-covered ground. This cold air is so dense that it pushes against Earth's surface with extreme pressure, preventing any moisture or heat from staying within the system.

Meteorologists have identified many semi-permanent areas of high-pressure. The Azores high, for instance, is a relatively stable region of high pressure around the Azores, an archipelago in the mid-Atlantic Ocean. The Azores high is responsible for arid temperatures of the Mediterranean basin, as well as summer heat waves in Western Europe.

Global Scale Meteorology

Global scale phenomena are weather patterns related to the transport of heat, wind, and moisture from the tropics to the poles. An important pattern is global atmospheric circulation, the large-scale movement of air that helps distribute thermal energy (heat) across the surface of the Earth.

Global atmospheric circulation is the fairly constant movement of winds across the globe. Winds develop as air masses move from areas of high pressure to areas of low pressure. Global atmospheric circulation is largely driven by Hadley cells. Hadley cells are tropical and equatorial convection patterns. Convection drives warm air high in the atmosphere, while cool, dense air pushes lower in a constant loop. Each loop is a Hadley cell.

Hadley cells determine the flow of trade winds, which meteorologists forecast. Businesses, especially those exporting products across oceans, pay close attention to the strength of trade winds because they help ships travel faster. Westerlies are winds that blow from the west in the midlatitudes. Closer to the Equator, trade winds blow from the northeast (north of the Equator) and the southeast (south of the Equator).

Meteorologists study long-term climate patterns that disrupt global atmospheric circulation. Meteorologists discovered the pattern of El Nino, for instance. El Niño involves ocean currents and trade winds across the Pacific Ocean. El Niño occurs roughly every five years, disrupting global atmospheric circulation and affecting local weather and economies from Australia to Peru.

El Niño is linked with changes in air pressure in the Pacific Ocean known as the Southern Oscillation. Air pressure drops over the eastern Pacific, near the coast of the Americas, while air pressure rises over the western Pacific, near the coasts of Australia and Indonesia. Trade winds weaken. Eastern Pacific nations experience extreme rainfall. Warm ocean currents reduce fish stocks, which depend on nutrient-rich upwelling of cold water to thrive. Western Pacific nations experience drought, devastating agricultural production.

Uses of Meteorology

Weather Forecasting

The most obvious public face of the science of meteorology is in weather forecasting. Every time we turn on the news to understand how our local weather is going to look for today or next few days, we are utilizing one of the most common applications of meteorology. It applies many scientific methods and tools in attempting to predict, through atmospheric indicators, how the weather conditions will look one hour, one day or one week from now. The further into the future, the more erratic and less predictable the conditions become.

Meteorologists who work in this area collect quantitative data (mathematical information and statistics) on such information as air pressure, present weather, wind in the different levels of the atmosphere to create forecast models based on recognition of past patterns, mitigation of model bias. Weather people often get predictions wrong because of something called "Chaos Theory" (the concept that conditions can change wildly based on minor fluctuations in conditions). Newtonian physics once determined that systems were stable, but Einstein determined they are erratic, unpredictable (to a degree) and subject to external influences based on minute changes. Multiple models are used today to increase accuracy and super-fast computational processes can highlight up-to-the-minute changes.

We rely on weather warnings - for tornados, hurricanes, flooding, heavy rain and snow etc. for safety advice, protecting our lives and homes too.

Commodity Trading

Perhaps one of the most surprising ways in which meteorology is applied is in commodities trading. Stocks and shares trade is one area of employment for meteorologists, especially when the dealing with commodity crops such as coffee (affected by adverse weather conditions) and fuel (we use more in unusually cold winters). These organizations trade based on longer-term weather forecasting and what a crop harvest is going to look like in a certain year. Thales of Miletus was perhaps the first person to do this when he predicted a bumper crop for olives, bought a press, and made a lot of money in the process. It's an inexact science, because weather conditions that will create a bumper harvest for one crop may prove destructive for another. There are many examples of how meteorology has presented speculators with opportunities to make money.

Even smaller businesses such as clothing retailers and restaurants are utilizing data from meteorological data specialists. Targeted advertising is set to go out at certain times such as commercials for wet weather clothing during unusually wet weather and sun cream during unusually warm weather, not just based on typical seasonal trends.

Aviation Meteorology

This division of meteorology deals with military and commercial flying and weather conditions in the upper levels of the atmosphere. Even when the weather is good at ground level, it doesn't mean the same conditions apply 30,000ft. Aviation meteorology is the applied science that dictates air traffic - whether a route is safe or dangerous, at what times, and whether flights can be made at all. They will disseminate data about head and tail-winds, temperature changes, ice buildup (which can damage aircraft performance) and variation on the ground, air pressure variation across the world and through the atmosphere, visibility and local conditions advisory systems for pilots. They can dictate when it's unsafe to take off or unsafe to land (and find alternate airports such as the eruption of the Icelandic volcano in that causes havoc across North America and Europe.

Agricultural Meteorology

Few industries and areas of our lives are as dependent on changes in weather conditions as agriculture. Crops for food and for clothing are necessary to live and for business, providing livelihoods for those who grow crops - not just food, but also commodity crops such as cotton and coffee. Meteorology determines when farmers should sow, when they should reap, and what steps they will need to take to protect crops from erratic weather. They may need to engage in flood mitigation or effective water management during drought to protect from crop failure.

Throughout the season from sowing to harvest, farmers and agricultural workers must engage in proper crop management and monitoring. This means effective watering or drainage but also includes ensuring the right nutrients remain in the soil for the crop and for the season but is also based their forecast crop yields on weather conditions and how quickly they may respond to changing patterns. Neither does meteorology just apply to crop management; livestock management for milk production depends on weather conditions too. Finally, some use agricultural meteorology as an applied science to determine the relationships between a local environment, crops, soil types, soil profile, and understanding which crops can and cannot grow in certain types of soil.

Environmental Meteorology

Environmental meteorology is concerned with the study of pollution and its effects on the climate as a forcing of local, regional and national weather patterns. It will look at such aspects as variation in temperatures, water vapor density (humidity), speed and intensity of wind, and many other weather conditions and phenomena. It will also look at the physics of meteorological processes of acoustical, electrical, optical, and the thermodynamic processes of the atmosphere. Cloud formation, precipitation, weather conditions and much more. Rather than looking at just the weather conditions resulting from meteorology, it also examines the potential impacts of weather conditions on the environment and on climate. Extreme weather can change a landscape significantly and therefore alter the weather patterns. Long-term and large-scale modelling, data accumulation and analyses and modelling feature heavily in environmental modelling.

Hydrometeorology

This is the subdivision of meteorology that fuses with the science of hydrology, examining how water transfers from dry land (evaporation) and lower echelons of the atmosphere. This examines cloud formation and the processes that lead to water-based natural phenomena, but it also predicts, projects and studies water hazards such as flooding and tropical cyclones, but also the effects of drought and land desertification - all these things concern precipitation and hydrology. Common applications behind this branch include "the water budget" - the account of all water transferring in or out of a water asset. Too much and it will flood, too little and drought ensues. Hydrometeorologists also monitor variation, quantity, intensity and distribution of rainfall. Examination of snowstorms also falls within the remit of hydrometeorology

This is multidisciplinary, using applied math, statistics, but also computer data modeling. Problems with the cross-disciplinary approach and lack of broad expertise are being addressed by some projects such as DRIHM which aims to use big data and broad methodology to improve hydrometeorological forecasting for the future and mitigate the effects of extreme weather. That EU project finished in 2015, but its results will be useful for decades to come.

Synoptic Meteorology

Rather like contour lines on a map, how close or far apart they are, determines the weather pattern. In cartography, these contours denote the steepness of the elevation. In meteorology, they serve a similar purpose - denoting density. Synoptic Meteorology examines large-scale weather systems and their impact on larger areas. It looks at the systems that form to create hurricanes and cyclones, including weather fronts, their directions and jet streams. Conditions must be right from a variety of directions and sources to determine whether or not a certain weather system will develop. It examines the structures of the atmosphere and its behaviors in attempting to forecast systems. Weather forecasting needs to look at the conditions way beyond the area of study to understand what the weather may be like in a certain area. Synoptic Meteorology, then, takes a broader view.

Maritime/Marine Meteorology

Where conditions on land are vital for agricultural workers, those who work at sea - industrial fishing, oil and gas rigs, military navy, commercial shipping, all need up-to-date information on the weather conditions where they are or where they are going to dictate operations. They need to take action to avoid or mitigate the effects of extreme weather at sea, such as storms and hurricanes, cyclones, icebergs, the jet stream and much more. For commercial fishing, the problems can go beyond merely avoiding terrible storms in the name of safety. Business decisions must be made while taking on board changes in sea and freshwater weather systems. Even after these conditions have passed, fish stocks may be higher or lower than they were prior to the storm. Such information is also required for leisure boat activities and public transportation across water bodies. This information is not just useful for people who work at sea; many weather fronts develop on the ocean and then hit land, affecting activities on the shore.

Military Meteorology

Armed forces all around the world rely heavily on weather conditions and meteorological forecasts to plan military operations and training exercises. Some of the biggest upsets in military history have come from adverse weather conditions. In World War II, the Nazi Regime troops were held back by an advancing Russian winter at Stalingrad. A century earlier, Napoleon had the same problem. The Spanish Armada's proposed invasion of England in 1588 failed because of storms in the channel - so did Julius Caesar's invasion of Britannia. On the flipside, the D-Day landings in France during WWII were meticulously planned and executed after lengthy examination of the weather conditions in the English Channel. Originally set for June 5th 1944, commanders postponed due to weather warnings from meteorologists. At the time of the postponement, the weather was calm, but it did turn bad around the originally proposed time of mission launch. No matter how well-trained the forces, military commanders the world over do not underestimate the impact that weather conditions can have.

Nuclear Meteorology

How do we detect radioactive particles in the atmosphere? How do we determine their impact on the environment? Radiation is a natural phenomenon; this young subdivision of meteorology arose during the 19th and 20th centuries as humanity began to understand radiation waves, their sources and effects, and of course the results of nuclear testing since the 1930s. Nuclear meteorology investigates the distribution of radioactive gases and aerosols. They are present at nuclear energy production facilities, monitoring the reactors for radiology leaks and predicting their effects on the environment. They work to ensure environmental compliance at any facility where nuclear technology is in use. Most of all, they monitor pollution distribution by examining air currents and turbulence and predicting where the material will spread. They were integral to examining the spread of nuclear material following the meltdown of Chernobyl and relaying reports to European governments about the disaster.

Renewable Energy

As supplies of fossil fuel run out, the world can expect to move increasingly towards renewable energy. In 2018, this is a growth area but has a long way to go to catch up with conventional energy use. Most renewable energy sources are highly dependent on weather conditions. In some cases, that begins from even before the resources are planned and built. Wind farms need to be placed in areas used experiencing high winds. Solar farms need to go in areas that get a lot of sunshine. Hydroelectric power requires regular, sustainable and predictable water sources. Meteorologists examine local weather system history to determine sites of best placement. They must also take into account freak, erratic and variation in weather. Calm weather reduces the capacity of wind turbines, overcast, cold and wet weather reduce solar panel effectiveness, and drought will reduce the effectiveness of hydroelectric systems. Biofuel such as elephant grass needs the right climate and weather conditions for a sustainable crop and long-term productivity. In cold weather conditions when demand is high, there is potential for production to be low depending on the methods used. Finally, errors in forecasting can lead to reduced accessibility and lost money for producers. Therefore, renewable energy planning and development is and will remain fundamental to renewable energy at every stage.

Extreme Weather and Disaster Management

Extreme weather is a fact of life. California's ongoing drought and forest fires continue to spread year on year. Hurricanes such as Katrina cause high winds and flood urban centers. In 2017, no country was untouched by natural disaster somewhere within its borders. Meteorology is useful in examining and planning the spread of extreme weather beforehand and offering advice before it happens, the science is just as important during and after. Disaster relief organizations such as FEMA need to understand weather conditions while they plan relief efforts. As with military meteorology above, the weather alone can be the difference between success and failure. When weather conditions are part of the plan with a disaster relief strategy, professionals are able to provide relief where they can and to do so safely without risking the lives of recipients or their own. Planning will often begin days in advance to see when the worst conditions have subsided and may be subject to change each day with fluctuating weather conditions.

Meteorological Methods

Persistence Forecast

This is the simplest method to use and comes in quite useful; it starts with an assumption that present conditions will not change. Based on season expectations and averages, the weather tomorrow will be much the same as it is today. This works best in areas where the weather is not volatile or where it changes slowly. Southern California is a good example of this where conditions rarely change and seasonal alterations are less volatile and incremental with little variation noticeable on a day-to-day basis. Ideal for short-term forecasting, its limits are usually exposed when unusual weather fronts move in. It is not particularly useful for long-range forecasting.

Trends Forecast

The trends forecasting method examines direction and speed of weather fronts, pressure bars, and cloud and precipitation build-up. This data is used to predict what the weather will be like in a certain area in a few hours or days based on what it is like now elsewhere. This relies on understanding of the conditions that cause a condition to intensify or dissipate as it progresses. This is best used to predict rain fronts. They will examine such elements as wind speed to predict when they will arrive. Weather is fairly predictable but can be subject to fluctuation based on the chaotic nature of new fronts forming and other forcings.

Numerical Weather Prediction

One of the most recent developments, it uses applied math to define weather conditions, patterns and trends. Today, meteorological organizations use computer modelling to on powerful computer systems to output a large number of predictions on all manner of atmospheric conditions. This hard data is then used to predict potential weather conditions short and long-term, and short and long-range. These supercomputers process thousands of computations every second to provide up-to-the-minute forecasts. They are not always right, but weather forecasts are right more often than not thanks to these computerized predictions. Often, errors are down to human error in inputting, insufficient data, and the chaotic nature of present weather conditions. When the equations are at fault, so will be the results. Other problems with the method include the lack of data from extreme environments. It's often difficult to acquire data from the middle of the ocean and tops of mountains, but satellite imagery can alleviate some of these problems.

Analog Method Forecasting

This is a method of comparison. In many ways, it is the opposite of Persistence Forecasting and works for some climate types more than others - particularly where erratic weather is common. Forecasters look to a date in the past (usually in the same season) to predict what the weather will be like tomorrow based on past experience. The assumption is that the change in weather pattern will mirror that of the past. If the weather is warm today but there is a change in wind direction or a cold front heading towards you, rather than assuming it will remain warm, forecasters will look for situations in the past where the same thing happened and attempt to predict how the weather might change. This can work well in predicting storms and other intense weather fronts. It has its problems, mostly because it relies on uniformity; if the weather has proven anything, it's that it is rarely uniform.

Climate-Based Methods

Our understanding of meteorological phenomena now has a new variable: climate change. We know that weather condition are changing globally based on carbon emissions. It's understood that a warming climate will not lead to uniform warming everywhere. Some will become warmer and wetter, some warmer and drier. Some will experience slowing of their ocean jetstreams and experience cooling - colder and wetter weather. Global climate change has the potential to change not just weather conditions, but change regional rules too. Erratic weather will become common and systems we have come to rely on to determine weather patterns may become near-useless. As the climate changes, meteorologists need to take seasonal averages over many years to understand what the weather might be like today or tomorrow or to predict long-term forecasts for future developments. This can also inform medical sciences and the spread of epidemics.

Climatology

Climatology is the scientific study of climate. Climate is defined as weather patterns that have been averaged over a given period of time to obtain a consistent pattern of the expected atmospheric conditions.

Climatology is the scientific study of climate. Climate is defined as weather patterns that have been averaged over a given period of time to obtain a consistent pattern of the expected atmospheric conditions. Weather is the atmospheric condition of a particular place over a short period of time, normally a day. Weather averaging for a long and indefinite period of time makes it possible to predict the climatic pattern of an area. Climatology is regarded as a subdivision of physical geography, atmospheric sciences, and earth sciences in general. Aspects of oceanography and biogeography have also been considered as part of climatology. Climatology focuses on aspects such as atmospheric boundary layer, circulation patterns, heat transfer in the globe, ocean interaction with the atmosphere and land surface, land use and topography.

Scientific Nature and Scope of Climatology

Climatology has evolved from a simple theoretical book keeping activity to the current complex scientific and a practical field. Science is defined as the truths and facts that have been obtained through constant research, systematic methods, evaluation of phenomena, and observation. Climatology is therefore scientific method that involves all the aspects that define science. Other than the above stated aspects, to obtain climatic patterns, several scales and gauges are employed in the climatic research. Climatology is not only concerned with the climate of a place but it also establishes the reason for the fluctuation of climate in the area, how human activities lead to climatic variations, effects of the climate on human activities, and the characteristics of the climate. Climate also depends on the layers of the earth and atmosphere, a further manifestation of its scientific nature.

Sub-Fields of Climatology

According to the area of specialization, climatology has been divided into smaller sub-fields: paleoclimatology, paleotempestology, historical climatology, metrology, and bioclimatology. Paleoclimatology focuses on establishing the past climatic patterns of a place by studying ice cores and tree rings. Paleotempestology uses ancient data to determine the frequency and magnitude of past hurricanes. Historical climatology focuses on establishing the climate of a place after studying the activities that the ancient dwellers of the particular place engaged in. Metrology, which is often confused with climatology, deals with weather, which runs for a maximum of probably a week or a month. Bioclimatology deals with the effects the climate has on the living organisms.

Importance of Climatology

Climatology is important in determining the climatic patterns of a particular region. Establishing the climatic pattern is significant in deciding the economic activities that would thrive in that particular region. If the climate of a region is established to be cool and wet, it would be safe to conclude that agriculture might thrive in the region. Having a clear climate pattern makes it easier for people to understand the seasons of engaging in particular tasks. This is especially most important to tourists and farmers. Infrastructure development, especially buildings are dependent on climate. After a climatic pattern has been established, the engineers recommend the use of materials that would not only withstand the conditions but also protect the dwellers from any harsh climatic condition. Furthermore, climatology seeks to establish why climate varies from place to another.

Urban Climatology and Climate Change

The understanding of global warming, or more accurately global climate change, as a result of human actions is a major research area in climatology. There are several specific branches of the science that inform our understanding of how humans are currently impacting the climate, but urban climatology has become a major focus in recent years.

Urban climatology is concerned with how urban growth affects the atmosphere and vice versa. The interest in city weather began in the early 19th century, but it wasn't until the latter part of the 20th century that the tools for truly understand the effects of large-scale city building on the climate became available.

It has long been known that cities, because of their use of dark asphalt and concrete surfaces, retain heat. It took over a century, however, to begin to understand how that retained heat, now known as the urban heat island (UHI) effect, impacts the weather and climate. Satellite data has been instrumental in the understanding that UHIs can affect local wind patterns, precipitation, cloud cover, fog, and humidity. Probably most profound is the fact that UHIs retain heat and that they lead to more of an increase in nighttime temperatures than daytime temperatures. The result is a lack of cooling during hot summer periods and thus increased use of air conditioning. Of course, air conditioning relies on fossil fuels for electricity generation, so even if the UHI does not directly affect the climate by changing weather patterns, it does lead to massive increases in fossil fuel use and thus increased greenhouse gas output.

Climate Indices and Models

The modern study of climatology, particularly dynamic climatology, relies on the understanding of climate indices. Climate indices are large-scale weather patterns that are consistent and measureable. The goal of an index is to combine a number of factors into a large, generalized description of either air or ocean phenomena that can be used to track the global climate system.

The indices used today include the following,:

- El Nino - Southern Oscillation (ENSO)

- Interdecadal Pacific Osciallation (IPO)

- Madden-Julian Oscillation (MJO)

- North Atlantic Oscillation (NAO)

- Northern Annular Mode/Arctic Oscillation (NAM/AO)

- Northern Pacific Index (NPI)

- Pacific Decadal Oscillation (PDO).

Each index above indicates a single, large-scale conglomeration of related weather patterns that have predictive value for general weather trends. Because of their predictive value, these indices, along with ice cover in the Arctic and Antarctic (and even information gleaned from studying the climates of other planets in the Solar System) are used to construct climate models.

The goal of a climate model is to project future trends in Earth's weather and climate. This includes things like overall average temperature, ocean acidity, trends in storm strength and frequency, etc. There are several different categories of climate model as follows:

- Zero-dimensional Models – These are simple models that take into account the amount of radiation entering and leave the atmosphere. The idea is that the two should be roughly equivalent otherwise the planet's temperature will change.

- One-dimensional Models – These take into account everything that a zero-dimensional model does, but also factor in the convective movement of heat through the atmosphere. These models are used to determine how changes in various greenhouse gas concentrations will affect surface temperature. A one-dimensional model is also called a radiative-convective model because it factors in the vertical movement of heat by convection in the atmosphere.

- Higher-dimensional Models – Adding parameters to a one-dimensional model leads to more complicated models with more predictive power. For instance, factoring in the horizontal movement of energy in the atmosphere or how albedo (reflection of sunlight by ice) affects energy transfer are both ways of increasing the dimensions of a climate model.

- Global Climate Models – Also called general circulation models, these models take into account the movement of energy in three dimensions as well as over time. Most of these models, despite being around since the late 1960s are considered "under development" and have areas of uncertainty.

Glaciology

Glaciology is the study of ice and its effects. Since ice can appear on or in the earth as well as in its seas and other bodies of water and even its atmosphere, the purview of glaciologists is potentially very large. For the most part, however, glaciologists' attention is direction toward great moving masses of ice called glaciers, and the intervals of geologic history when glaciers and related ice masses covered relatively large areas on Earth. These intervals are known as ice ages, the most recent of which ended on the eve of human civilization's beginnings, just 11,000 years ago. The last ice age may not even be over, to judge from the presence of large ice masses on Earth, including the vast ice sheet that covers Antarctica. On the other hand, evidence gathered from the late twentieth century onward indicates the possibility of global warming brought about by human activity.

Working of Glaciology

Ice

Ice, of course, is simply frozen water, and though it might appear to be a simple subject, it is not. Glaciologists classify differing types of ice, for instance, with regard to their levels of density, designating them with Roman numerals. The ice to which most of us are accustomed is classified as ice I. We will not be concerned with the other varieties of ice in the present context, but it should be noted that the ice in glaciers is quite different from the ice in an ice cube or even the ice on a pond in winter. These differences are a result of massive pressure, which reduces the air content of the ice in glaciers.

By definition, ice is composed of fresh water rather than saltwater. This is true even of icebergs, though they may float on the salty oceans. The reason is that water has a much higher freezing point than salt, and, therefore, when water freezes, very little of the salt remains joined to the water. Most of the salt is left behind in the form of a briny slush, and so much of Earth's fresh water supply actually is contained in great masses of ice, such as the glaciers of Antarctica.

Glaciology is defined as the study of ice, its forms, and its effects. This means that the glaciologist has a much wider scope than a geologist, meteorologist, or oceanographer, each of whom is concerned primarily with the geosphere, atmosphere, and hydrosphere, respectively. Though ice commonly is associated with the hydrosphere, where it appears on Earth's oceans, rivers, and lakes, it also is found on and even under the solid earth. There are even situations in which ice is found in the atmosphere.

Glaciology and Glaciers

Despite the wide distribution of ice on Earth and the many forms it takes, the work of most glaciologists is concerned primarily with ice as it appears in glaciers. A glacier is a large, typically moving mass of ice on or adjacent to a land surface. It does not flow, as water does; rather, it is moved by gravity, a consequence of its extraordinary weight.

Obviously, a glacier can form only in an extremely cold region—one so cold that the temperature never becomes warm enough for snow to melt completely. Some snow may melt as a result of contact with the ground, which is likely to be warmer than the snow itself, but when temperatures drop, it refreezes. A glacier starts with a layer of ice, on which snow gathers until refreezing gradually creates compacted layers of snow and ice.

As anyone who has ever held a snowball in his or her hand knows, snow is fluffy, or, to put it in more scientific terms, it is much less dense than ice. A sample of snow is about 80% air, but as ice accumulates over a layer of snow, the weight of the ice squeezes out most of the air. As the layers grow thicker and thicker, the weight reduces the air further, creating an extremely dense, thick layer of ice. Ultimately, the ice becomes so heavy that its weight begins to pull it downhill, at which point it becomes a glacier.

Glacial Temperature and Morphologic Characteristics

Glaciers may be classified according to either relative temperature or morphologic characteristics (i.e., in terms of its shape). In terms of temperature, a glacier may be "warm," meaning that it is close to the pressure melting point. Pressure melting point is defined as the temperature at which ice begins to melt under a given amount of pressure. It is commonly known that water melts at 32°F (0°C), but only under conditions of ordinary atmospheric pressure at sea level. At higher pressures, the melting point of water is lower, which means that it can remain liquid at temperatures below its ordinary freezing point. (The melting point and freezing point of a substance are always the same.)

A "warm" glacier, such as those that appear in the Alps, is relatively mobile, because it is at the pressure melting point. This kind of glacier contrasts with a "cold," or polar, glacier, in which the temperatures are well below the pressure melting point; in other words, despite the extremely high pressure, the temperature is so low that the ice will not melt. As their name suggests, polar glaciers are found at Earth's poles, which effectively means Antarctica, since the area of the North Pole is not a land surface. A third category of glacier, in terms of temperature, is a subpolar glacier, found (not surprisingly) in regions near the poles. Examples of subpolar glaciers, or ones in which the fringes of the glacier are colder than the interior, are found in Spits-bergen, islands belonging to Norway that sit in the Arctic Ocean, well to the north of Scandinavia.

Morphologic Classifications

In the classification of geologic sciences, glaciology often is grouped with geomorphology. The latter field of study is devoted to landforms, or notable topographical features, and the forces and processes that have shaped them. Among those forces and processes are glaciers, which can be viewed in terms of their shape, the locale in which they form, and their effect on the contour of the land.

Alpine or mountain glaciers flow down a valley from a high mountainous region, typically following a path carved out by rivers or melting snow in warmer periods. They move toward valleys or the ocean, and in the process they exert considerable impact on the surrounding mountains, increasing the sharpness and steepness of these landforms. The rugged terrain in the vicinity of the Himalayas and the Andes, as well as the alpine regions of the Cascade Range and Rocky Mountains in the United States, are partly the result of weathering caused by these glaciers.

The glacial forms found in Alaska, Greenland, Iceland, and Antarctica are often piedmont glaciers, large mounds of ice that slope gently. Iceland, Greenland, and Antarctica as well as Norway are also home to cirque glaciers, which are relatively small and wide in proportion to their length. Though they experience considerable movement in place, they usually do not move out of the basinlike areas in which they are formed.

Other Ice Formations

There are several other significant varieties of ice formation, including ice caps, ice fields, and ice sheets. An ice cap, though much bigger than a glacier, typically has an area of less than 19,300 sq. mi. (50,000 sq km). Nonetheless, its mass is such that it exerts enormous weight on the land surface, and this exertion of force allows it to flow.

At the center of an ice cap or an ice sheet is an ice dome, and at the edges are ice shelves and outlet glaciers. Symmetrical and convex (i.e., like the outside of a bowl), an ice dome is a mass of ice often thicker than 9,800 ft. (3,000 m). An outlet glacier is a rapidly moving stream of ice that extends from an ice dome. Ice shelves, at the far outer edges, extend into the oceans, typically ending in cliffs as high as 98 ft. (30 m). Ice fields are similar to ice caps; the main difference is that the ice field is nearly level and lacks an ice dome. There are enormous variations in size for ice fields. Some may be no larger than 1.9 sq. mi. (5 sq km), while at different times in Earth's history, others have been as large as continents.

The most physically impressive of all ice formations, an ice sheet is a vast expanse of ice that gradually moves outward from its center. Ice sheets are usually at least 19,300 sq. mi. (50,000 sq km) and, like ice caps, consist of ice domes and outlet glaciers, with outlying ice shelves. Given their even greater size compared with ice caps, ice sheets exert still more force on the solid earth beneath them. They cause the rock underneath to compress, and, therefore, if an ice sheet ever melts, Earth's crust actually will rise upward in that area.

Types of Glaciation

Glaciologists, who focus on the study of glaciers, have identified two types of glaciation that occurs on the earth's surface: alpine and continental.

- Alpine glaciation refers to accumulations of ice that is confined to valleys. This ice moves down these valleys via mountainsides in order to reach the large plain areas below. During their slow movement, alpine glaciers carve out the landscape, deepening ravines and sharpening ridge lines. As a result, the land near alpine glaciers appear to be more rough and jagged.

- Continental glaciation refers to large areas of ice coverage, which has accumulated over a wide range of landscapes and is mainly found in the northern hemisphere. Continental glaciers cover thousands of square miles and measure thousands of feet in thickness. Today, they are mainly found in Greenland and Antarctica. These glaciers tend to wear down the surrounding landscape into smooth surfaces.

Types of Glacial Deposits

Glaciers move in a number of ways and at a variety of speeds. These factors work together to influence the types of deposits left behind by these moving masses of ice. The two major categories of glacial deposits are stratified and unstratified.

- Stratified glacial deposits are divided into 5 subcategories: kames, eskers, varves, kettles, outwash sand and gravel. Kames deposits are layers of drift carried by glaciers. This drift is deposited along the sides of steep, low elevation hills. Esker deposits are ridges of gravel and sand found along steep inclines. These deposits have been left behind by water moving under ice. Varve deposits are found within proglacial lakes. These are changing layers of deposit that alternate between course (from the summer season) and fine (from the winter season) sediment. Kettle deposits occur when a large body of ice forms a hollowed out area in the ground. Outwash sand and gravel refers to what is left behind on flat areas and plains. This sediment was deposited from the front of a moving glacier.

- Unstratified deposits are divided into 4 subcategories: moraines, ribbed-moraines, drumlins, and till-unsorted. Moraine deposits refer to the sediment and waste left behind as glaciers melt. These deposits may occur on all sides of the glacier. Ribbed-moraine deposits create long, subglacial hills. Drumlin deposits are smooth hills that are made up of till. Till-unsorted deposits occur in a range of sediment size, from flour-like to boulder-like. Till-unsorted deposits may form both moraines or drumlins.

Oceanography

Oceanography is the study of all aspects of the ocean. Oceanography covers a wide range of topics, from marine life and ecosystems, to currents and waves, to the movement of sediments, to seafloor geology.

The study of oceanography is interdisciplinary. The ocean's properties and processes function together and cannot be examined separately from one another. The chemical composition of water, for example, influences what types of organisms live there. In turn, organisms provide sediments to the geology of the seafloor. Oceanographers must have a broad understanding of these relationships to research specific topics, or subdisciplines.

Subdisciplines of Oceanography

Oceanography's diverse topics of study are generally categorized in four major subdisciplines. A subdiscipline is a specialized field of study within a broader subject or discipline. Oceanographers specialize in the biological, physical, geological, and chemical processes of the marine environment.

Biological Oceanography

Biological oceanographers study how each of the subdisciplines of oceanography work either separately or together to influence the distribution and abundance of marine plants and animals, as well as how marine organisms behave and develop in relation to their environment. Marine biologists and fisheries scientists are biological oceanographers.

Biological oceanographers also focus on how species adapt to environmental changes, such as increased pollution, warming waters, and natural and artificial disturbances. A natural disturbance may be the eruption of an underwater volcano or a hurricane, while an artificial disturbance may be an oil spill or overfishing.

The Cetacean Sanctuary Research Project is a marine biology program that focuses on whale and dolphin species (cetaceans) living in the Pelagos Sanctuary in the northwestern Mediterranean Sea. A range of human activities threatens these species—intense maritime traffic, urban pollutants, and oil and gas exploration. By analyzing cetacean behavior in this high-pressure environment, oceanographers hope to protect Pelagos' marine species and promote their importance to the surrounding coastal community.

To get a clear picture of these species and their behavior, oceanographers monitor their geographic position, movements, and group size. They record vocalizations, respiration patterns, and surface and aerial displays to understand how cetaceans interact with one another and other marine species. Skin and fecal samples are analyzed to generate information on social, sexual, and feeding habits. The Cetacean Sanctuary Research Project has collected one the largest data sets on marine mammals in the Mediterranean Sea.

New technology is expanding opportunities for biological oceanographers. Marine biotechnology is the study of how marine resources can be used to develop industrial, medical, and ecological products.

Using a process called biomimicry, researchers are able to understand, isolate, and fabricate biological properties of marine species. Natural compounds found in corals and other marine organisms have potent anti-cancer properties. Proteins found in marine algae and bacteria are being developed into super-absorbent materials that could be used to clean oil spills. Salt-tolerant land crops have been created from genetically engineered marsh plants.

The real-world applications of this research are potentially endless because marine species make up more than 80 percent of Earth's living organisms.

Physical Oceanography

Physical oceanographers study the relationship between the ocean's physical properties, the atmosphere, and the seafloor and coast. They investigate ocean temperature, density, waves, tides, and currents. They also focus on how the ocean interacts with Earth's atmosphere to produce our weather and climate systems.

Oceanographers in South Africa, for example, have studied the turbulent flow of water around the southern tip of Africa. This movement, known as the Agulhas Current, is part of a larger "ocean conveyor belt" that circulates water around the globe based on density, wind, and currents. Physical oceanographers have found that the amount of water flowing from the Indian Ocean to the Atlantic Ocean has increased, a process nicknamed the Agulhas leakage. The increased Agulhas leakage has been linked to global warming.

Physical oceanographers predict that global warming will slow the ocean conveyor belt and radically change climate and weather patterns. As ice caps melt, sea levels rise and the ocean becomes less salty and dense. As ocean waters warm, they also expand, which enhances sea level rise.

Geological Oceanography

Geological oceanographers study the past and present composition of seafloor structures. They investigate the origins of the underwater landscape and detail its development and changes. They also focus on the physical and chemical properties of rocks and sediments found on the seafloor.

A variety of geological research projects have been conducted on the JOIDES Resolution, an international research vessel. Resolution drills sediment-core samples and collects measurements from under the ocean floor. This research helps scientists understand our paleoclimate. Paleoclimatology is the study of weather and climate patterns over hundreds of millions of years. Changes in the seafloor reflect changes in Earth's climate, and is very useful in predicting our future climate.

Starting in December 2010, oceanographers and other scientists aboard the Resolution studied the Louisville Seamount Trail, a series of underwater volcanoes found in the South Pacific close to New Zealand. The vessel drilled sediment-cores at five different sites to understand the development of the hot spot that created these volcanoes. The results of this research will help define how landforms develop and change.

Chemical Oceanography

Chemical oceanographers study the chemical composition of seawater and its resulting effects on marine organisms, the atmosphere, and the seafloor. They map chemicals found in seawater to understand how ocean currents move water around the globe—the ocean conveyor belt. Chemical oceanographers study how the carbon from carbon dioxide is buried in the seafloor, highlighting the key role that the ocean plays in regulating greenhouse gases, such as carbon dioxide, that is a major contributor to global warming. Chemical oceanographers also focus on how pollutants affect seawater composition. They may study the unusual and sometimes toxic fluids released by hydrothermal vents in the ocean floor.

Ocean acidification is a key topic in chemical oceanography. The ocean is becoming more acidic because of the increased amount of carbon dioxide in the atmosphere. Acid disrupts the formation of calcium carbonate, the basic building block of shells and corals.

Shellfish populations in the Pacific Northwest region of the U.S. have declined dramatically as a result of ocean acidification. Chemical oceanographers in Oregon help shellfish growers adjust their operations to reduce the influx of acidic water. They also run experiments to find the threshold at which shellfish are unharmed by acidification. This research will complement other studies that aim to reduce the negative impacts of ocean acidification in shellfish and coral environments around the globe.

Ocean Acidification

Ocean acidification describes the decrease in ocean pH that is caused by anthropogenic carbon dioxide (CO_2) emissions into the atmosphere. Seawater is slightly alkaline and had a preindustrial pH of about 8.2. More recently, anthropogenic activities have steadily increased the carbon dioxide content of the atmosphere; about 30–40% of the added CO_2 is absorbed by the oceans, forming carbonic acid and lowering the pH (now below 8.1) through ocean acidification. The pH is expected to reach 7.7 by the year 2100.

An important element for the skeletons of marine animals is calcium, but calcium carbonate becomes more soluble with pressure, so carbonate shells and skeletons dissolve below the carbonate compensation depth. Calcium carbonate becomes more soluble at lower pH, so ocean acidification is likely to affect marine organisms with calcareous shells, such as oysters, clams, sea urchins and corals, and the carbonate compensation depth will rise closer to the sea surface. Affected planktonic organisms will include pteropods, coccolithophorids and foraminifera, all important in the food chain. In tropical regions, corals are likely to be severely affected as they become less able to build their calcium carbonate skeletons, in turn adversely impacting other reef dwellers.

The current rate of ocean chemistry change seems to be unprecedented in Earth's geological history, making it unclear how well marine ecosystems will adapt to the shifting conditions of the near future. Of particular concern is the manner in which the combination of acidification with the expected additional stressors of higher temperatures and lower oxygen levels will impact the seas.

Ocean Currents

Since the early ocean expeditions in oceanography, a major interest was the study of the ocean currents and temperature measurements. The tides, the Coriolis effect, changes in direction and strength of wind, salinity and temperature are the main factors determining ocean currents. The thermohaline circulation (THC) *thermo-* referring to temperature and *-haline* referring to salt content connects 4 of 5 ocean basins and is primarily dependent on the density of sea water. Ocean currents such as the Gulf Stream are wind-driven surface currents.

Ocean Heat Content

Oceanic heat content (OHC) refers to the heat stored in the ocean. The changes in the ocean heat play an important role in sea level rise, because of thermal expansion. Ocean warming accounts for 90% of the energy accumulation from global warming between 1971 and 2010.

Coastal Geography

The coast is a special environment due to its complex and dynamic construction. Three major elements of the nature - terrestrial, aquatic and atmospheric - meet on coastal areas. Coasts are also important areas of human settlement and activity. Coastal geography aims to describe and understand the interacting processes. Thus, the coastal zones are globally interesting environments for geographical research.

The natural and human processes and activities that are at work on the world's coasts induce great changes in the coastal environments. Understanding the individual processes and their site-specific form of interaction is a challenge for coastal geography. Gaining knowledge about the coastal systems enables assessment of the impacts of the global environmental change. The scale of factors that alter coastal areas range from local to global, even astronomic (tidal activity).

Coastal Zone

The coast is not a sharp border between land and sea, but a relatively wide transitional zone between them. Definitions of coast usually include land areas that are affected by marine influence and sea areas that are affected by terrestrial influence. Hence, it is difficult to precisely delineate the coast. Coasts are affected by different natural processes, such as shore processes, fluvial activity and aeolian processes. Matter and energy are transported over and along the shore. These factors make the coasts highly dynamic systems, which are sensitive to changes in any of the processes.

Palaeogeography

Paleogeography is also spelled palaeogeography is the ancient geography of Earth's surface.

Earth's geography is constantly changing: continents move as a result of plate tectonic interactions; mountain ranges are thrust up and erode; and sea levels rise and fall as the volume of the ocean basins change. These geographic changes can be traced through the study of the rock and fossil record, and data can be used to create paleogeographic maps, which illustrate how the continents have moved and how the past locations of mountains, lowlands, shallow seas, and deep ocean basins have changed.

The study of paleogeography has two principal goals. The first is to map the past positions of the continents and ocean basins, and the second is to illustrate Earth's changing geographic features through time.

Mapping Past Continents and Oceans

The past positions of the continents can be determined by using six major lines of evidence: paleomagnetism, linear magnetic anomalies, hot-spot tracks, paleobiogeography, paleoclimatology, and geologic and tectonic history.

Paleomagnetism

By measuring the remanent magnetic field often preserved in rocks containing iron-bearing minerals, paleomagnetic analysis can determine whether a rock was magnetized near one of Earth's poles or near the Equator. Iron-bearing minerals forming in igneous rock align themselves with Earth's magnetic field as the molten rock cools. These minerals also align themselves when they are deposited in sediments, and they retain their orientation as they lithify into sedimentary rock. Lines of force in Earth's magnetic field are parallel to the planet's surface at the Equator and are vertical at the poles. Therefore, iron-bearing minerals formed or deposited at low latitudes will be nearly parallel to Earth's surface, while those at high latitudes will dip steeply. If the rocks are later transported by tectonic processes, their original latitude of deposition can be determined by their orientation. Paleomagnetism provides direct evidence of a continent's past north-south (latitudinal) position, but it does not constrain its east-west (longitudinal) position.

Linear Magnetic Anomalies

Earth's magnetic field has another important property. Like the Sun's magnetic field, Earth's magnetic field periodically "flips," or reverses polarity—that is, the North and South poles switch places. Fluctuations, or anomalies in the intensity of the magnetic field, occur at the boundaries between normally magnetized sea floor and sea floor magnetized in the reversed direction. The age of these magnetic anomalies can be established by using fossil evidence and radiometric age determinations. Because these magnetic anomalies form at oceanic ridges, they tend to be long, linear features (hence the name linear magnetic anomalies) that are symmetrically disposed about ridge axes. The past positions of the continents during the last 150 million years (the maximum age of most of the ocean floor) can be directly reconstructed by superimposing linear magnetic anomalies of the same age, in effect "undoing" the results of sea-floor spreading since that time.

Hot-spot Tracks

Some of the world's volcanoes are formed by jets of molten rock that arise at the boundary between Earth's core and mantle (at a depth of about 2,900 km, or 1,800 miles). These rising plumes, or hot spots, puncture the lithosphere, and, as a tectonic plate moves across the hot spot, a line of islands is generated. The island directly above the hot spot is the youngest, and islands become progressively older with distance from the hot spot. There are more than a dozen well-documented hot-spot tracks. Perhaps the most obvious is the Hawaiian Islands, which trace an east-west arc across the central Pacific Ocean. Hot-spot tracks accurately record plate motions and can be used to determine the past latitudinal and longitudinal position of the continents.

Paleobiogeography

The past distribution of plants and animals can give important clues about the latitudinal position of the continents as well as their relative positions. Cold-water faunas can often be distinguished from warm-water faunas, and ancient floras reflect both paleotemperature and paleorainfall. The diversity of plants and animals tends to increase toward the Equator, and the adaptations of plants (such as smooth-edged leaves in the tropics and serrated-edged leaves in the temperate belts) are often good indicators of the amount of ancient rainfall.

The similarity or dissimilarity of faunas and floras on different continents can also be used to estimate their geographic proximity. In addition, the evolutionary history of groups of plants and animals on different continents can reveal when these continents were connected or isolated from each other. For example, Australia's unique marsupial fauna is the result of its isolation from the other continents at the time when placental mammals were evolving on the other continents during the early Paleogene Period.

Paleoclimatology

Earth's climate is primarily a result of the redistribution of the Sun's energy across the surface of the globe. It is warm near the Equator and cool near the poles. Wetness or rainfall also varies systematically from the Equator to the pole in alternating bands. It is wet near the Equator, dry in the subtropics, wet in the temperate belts, and dry near the poles. Certain kinds of rocks form under specific climatic conditions. For example, coals occur where wet climates once supported lush vegetation; bauxite (the principal ore of aluminum) is formed in warm and wet conditions, evaporites and calcretes require warmth and aridity to form; and tillites are deposited during the movement of glacial ice. The ancient distribution of these and other rock types can indicate how the global climate has changed through time and how the continents have traveled across the climatic belts.

Geologic and Tectonic History

In order to reconstruct the past positions of the continents, it is necessary to understand the evolution of plate tectonic boundaries. Only by understanding the regional geologic and tectonic history of an area can the location and timing of rifting events, subduction activity, continental collision, and other major plate tectonic events be determined.

Mapping Past Geographic Features

Paleogeographic features include mountain ranges, lowlands, shallow seas, and the deep ocean basins. Some paleogeographic features change very slowly and are easy to map. Others change very rapidly, so that any paleogeographic mapping is, at best, an approximation.

Continents and Ocean Basins

The two major paleogeographic features are the continents and the ocean basins. Since early Precambrian time, Earth has been divided into deep ocean basins (average depth, 3.5 km, or 2.2 miles) and high-standing continents (average elevation, about 800 metres, or 2,600 feet). Continental lithosphere stands high above the ocean basins because it is less dense and is not easily subducted, or recycled back into Earth's interior. Consequently, continents are made up of very old rocks, some dating back over 4 billion years. The amount of continental lithosphere has probably changed very little during the last 2.6 billion years—possibly increasing 10 to 15 percent. What has changed is the shape and the distribution of continents across the globe.

The ocean basins are also ancient paleogeographic features. Oceanic lithosphere is continuously created at oceanic ridges and then recycled back into Earth's interior at subduction zones.

Mountain Ranges

In contrast to the continents and ocean basins, which are permanent geographic features, the height and location of mountain belts constantly change. Mountain belts form either where oceanic lithosphere is subducted beneath the margin of a continent, giving rise to a linear range of mountains such as the Andes of western South America, or where continents collide, forming high mountains and broad plateaus such as the Himalayas and the Tibetan Plateau of Asia. Less-extensive mountains can also form when continents rift apart (as is happening today in the East African Rift) or where hot spots form volcanic uplifts.

In most cases, mountain ranges take tens of millions of years to form and, depending on the climate, may last for hundreds of millions of years. Though the Appalachian Mountains of the eastern United States were formed more than 300 million years ago, as a result of the collision of North America and western Africa, remnants of this collisional mountain belt still reach heights of over than 2,000 metres (6,600 feet). The Himalayan mountains, the world's tallest mountain range, began to rise from the sea nearly 50 million years ago when northern India collided with Eurasia.

Shorelines and Continental Margins

In contrast to mountain ranges, which take tens to hundreds of millions of years to uplift and erode, the location of Earth's shorelines can change rapidly. The familiar shapes that characterize today's shorelines such as Hudson's Bay, the Florida peninsula, or the numerous fiords of Norway are all less than 12,000 years old. The shape of the modern coastlines is the result of a rise in sea level of 70 metres (230 feet) that took place in the last 12,000 years as the last great ice sheet that covered much of North America and Europe melted.

It is important to note that the shoreline, though the edge of land, is not the edge of the continent. In most cases the continent extends seaward hundreds of kilometres beyond the shoreline. The actual edge of the continent in most cases is marked by the transition from the continental slope to the continental rise. This steep bathymetric gradient marks the boundary between continental and oceanic crust.

Agents Of Paleogeographic Change

The ancient distribution of land and sea, probably the single most important aspect of paleogeography, is a function of both continental topography and sea-level change. Though topography changes slowly (over tens of millions of years), global sea level can change rapidly (over tens of thousand of years). When sea level rises, the continents are flooded and shorelines move landward. Throughout much of Earth's history, sea level was higher than it is today, and vast areas of the continents were flooded by shallow seas.

Several factors can affect sea-level change. One factor is the amount of ice on the continents. At times when the continents are covered by great ice sheets, sea level is low, and the continents are more exposed. The last glacial maximum was 18,000 years ago. Other important global episodes of glaciation occurred 300 million, 450 million, and 650 million years ago. The oldest known glacial episode occurred in the Precambrian, approximately 2.2 billion years ago. For the last 20 million years, the continents and their margins have been largely high and dry because there has been a significant amount of ice on Antarctica and there has been extensive mountain-building in Asia.

Sea level also changes more slowly (over tens of millions of years) owing to changes in the volume of the ocean basins. During the Precambrian, gases escaped from Earth's interior and contributed to the formation of water vapour in the atmosphere. The vapour eventually condensed on the cooling surface to form the world's oceans. However, there has been no significant addition to the volume of water on Earth since early Precambrian times. Changes in sea level, therefore, are due not to changes in the amount of water on Earth but rather to changes in the shape and volume of the ocean basins. Plate tectonics and, in particular, sea floor spreading control the shape and volume of the ocean basins.

Environmental Geography

Environmental geography describes the spatial aspects of interactions between humans and the natural world. It requires an understanding of the dynamics of climatology, hydrology, biogeography, geology and geomorphology, as well as the ways in which human societies conceptualize the environment.

Environmental geography represents a critically important set of analytical tools for assessing the impact of human presence on the environment by measuring the result of human activity on natural landforms and cycles.

References

- Rice, A. L. (1999). "The Challenger Expedition". Understanding the Oceans: Marine Science in the Wake of HMS Challenger. Routledge. pp. 27–48. ISBN 978-1-85728-705-9

- Hydrology: newworldencyclopedia.org, Retrieved 11 April 2018

- Zeebe, R. E.; Zachos, J. C.; Caldeira, K.; Tyrrell, T. (4 July 2008). "OCEANS: Carbon Emissions and Acidification". Science. 321 (5885): 51–52. doi:10.1126/science.1159124. PMID 18599765

- Meteorology: environmentalscience.org, Retrieved 29 May 2018

- "Ocean acidification". Department of Sustainability, Environment, Water, Population & Communities: Australian Antarctic Division. 28 September 2007. Retrieved 17 April 2013

- Oceanography, encyclopedia: nationalgeographic.org, Retrieved 11 July 2018

- Gattuso, J.-P.; Hansson, L. (15 September 2011). Ocean Acidification. Oxford University Press. ISBN 978-0-19-959109-1. OCLC 730413873

- Paleogeography, science: britannica.com, Retrieved 28 June 2018

- Orr, James C.; et al. (2005). "Anthropogenic ocean acidification over the twenty-first century and its impact on calcifying organisms" (PDF). Nature. 437 (7059): 681–686. Bibcode:2005Natur.437..681O. doi:10.1038/nature04095. PMID 16193043. Archived from the original (PDF) on 25 June 2008

- Biogeography: newworldencyclopedia.org, Retrieved 31 March 2018

- "Sir John Murray (1841–1914) – Founder Of Modern Oceanography". Science and Engineering at The University of Edinburgh. Archived from the original on 28 May 2013. Retrieved 7 November 2013

3

Tools used in Physical Geography

A number of specialized tools and technologies are used in modern physical geography to understand, describe and explain the processes and structures of the Earth. The aim of this chapter is to explore the varied tools and techniques used in physical geography, such as GIS, GPS, remote sensing and maps.

Remote Sensing

Remote sensing is the process of acquiring details about an object without a physical on-site observation by means of satellite technology. Remote sensing is the process of acquiring details about an object without physical on-site observation using satellite or aircraft. Remote sensors are mounted on the aircraft or satellites to gather data via detecting energy reflected from the Earth. Remote sensing has been beneficial to scientists who are in constant need of data as it pertains to land, ocean, and the atmosphere.

Methods of Remote Sensing

Synthetic aperture radar image of Death Valley
colored using polarimetry.

Aerial and Satellite Imagery

Besides their role in photogrammetry, aerial and satellite imagery can be used to identify and delineate terrain features and more general land-cover features. These types of images increasingly have become part of geovisualization, whether as maps or GIS depictions. False-color and non-visible spectra imaging can also help determine the lie of the land by delineating vegetation and other land-use information more clearly. Images can be in visible colors and in other spectra.

Photogrammetry

Photogrammetry is a measurement technique for which the coordinates of the points of a multi-dimensional object are determined by measurements made in two photographic images (or more) taken starting from different positions, usually from different passes of an aerial photography flight. In this technique, the common points are identified on each image. A line of sight (or ray) can be derived from the camera location to the point on the object. The intersection of these rays (triangulation) determines the relative three-dimensional position of the point. Known control points can be used to give these relative positions absolute values. More sophisticated algorithms can exploit other information on the scene already known.

Radar and Sonar

Satellite radar mapping is one of the major techniques of generating Digital Elevation Models. Seismographic information can be useful in mapping sub-surface structures. Similar techniques are applied in bathymetric surveys using sonar or depth soundings to determine the terrain of the ocean floor.

Types of Data Acquisition Techniques

The basis for multispectral collection and analysis is that of examined areas or objects that reflect or emit radiation that stand out from surrounding areas. For a summary of major remote sensing satellite systems.

Applications of Remote Sensing

- Conventional radar is mostly associated with aerial traffic control, early warning, and certain large scale meteorological data. Doppler radar is used by local law enforcements' monitoring of speed limits and in enhanced meteorological collection such as wind speed and direction within weather systems in addition to precipitation location and intensity. Other types of active collection includes plasmas in the ionosphere. Interferometric synthetic aperture radar is used to produce precise digital elevation models of large scale terrain.

- Laser and radar altimeters on satellites have provided a wide range of data. By measuring the bulges of water caused by gravity, they map features on the seafloor to a resolution of a mile or so. By measuring the height and wavelength of ocean waves, the altimeters measure wind speeds and direction, and surface ocean currents and directions.

- Ultrasound (acoustic) and radar tide gauges measure sea level, tides and wave direction in coastal and offshore tide gauges.

- Light detection and ranging (LIDAR) is well known in examples of weapon ranging, laser illuminated homing of projectiles. LIDAR is used to detect and measure the concentration of various chemicals in the atmosphere, while airborne LIDAR can be used to measure heights of objects and features on the ground more accurately than with radar technology. Vegetation remote sensing is a principal application of LIDAR.

- Radiometers and photometers are the most common instrument in use, collecting reflected and emitted radiation in a wide range of frequencies. The most common are visible and infrared sensors, followed by microwave, gamma ray and rarely, ultraviolet. They may also be used to detect the emission spectra of various chemicals, providing data on chemical concentrations in the atmosphere.

- Spectropolarimetric Imaging has been reported to be useful for target tracking purposes by researchers at the U.S. Army Research Laboratory. They determined that manmade items possess polarimetric signatures that are not found in natural objects. These conclusions were drawn from the imaging of military trucks, like the Humvee, and trailers with their acousto-optic tunable filter dual hyperspectral and spectropolarimetric VNIR Spectropolarimetric Imager.

- Stereographic pairs of aerial photographs have often been used to make topographic maps by imagery and terrain analysts in trafficability and highway departments for potential routes, in addition to modelling terrestrial habitat features.

- Simultaneous multi-spectral platforms such as Landsat have been in use since the 1970s. These thematic mappers take images in multiple wavelengths of electro-magnetic radiation (multi-spectral) and are usually found on Earth observation satellites, including (for example) the Landsat program or the IKONOS satellite. Maps of land cover and land use from thematic mapping can be used to prospect for minerals, detect or monitor land usage, detect invasive vegetation, deforestation, and examine the health of indigenous plants and crops, including entire farming regions or forests. Landsat images are used by regulatory agencies such as KYDOW to indicate water quality parameters including Secchi depth, chlorophyll a density and total phosphorus content. Weather satellites are used in meteorology and climatology.

- Hyperspectral imaging produces an image where each pixel has full spectral information with imaging narrow spectral bands over a contiguous spectral range. Hyperspectral imagers are used in various applications including mineralogy, biology, defence, and environmental measurements.

- Within the scope of the combat against desertification, remote sensing allows researchers to follow up and monitor risk areas in the long term, to determine desertification factors, to support decision-makers in defining relevant measures of environmental management, and to assess their impacts.

Geodetic

- Geodetic remote sensing can be gravimetric or geometric. Overhead gravity data collection was first used in aerial submarine detection. This data revealed minute perturbations in the Earth's gravitational field that may be used to determine changes in the mass distribution of the Earth, which in turn may be used for geophysical studies, as in GRACE. Geometric remote sensing includes position and deformation imaging using InSAR, LIDAR, etc.

Acoustic and Near-acoustic

- Sonar: passive sonar, listening for the sound made by another object (a vessel, a whale etc.); active sonar, emitting pulses of sounds and listening for echoes, used for detecting, ranging and measurements of underwater objects and terrain.

- Seismograms taken at different locations can locate and measure earthquakes (after they occur) by comparing the relative intensity and precise timings.

- Ultrasound: Ultrasound sensors, that emit high frequency pulses and listening for echoes, used for detecting water waves and water level, as in tide gauges or for towing tanks.

- To coordinate a series of large-scale observations, most sensing systems depend on the following: platform location and the orientation of the sensor. High-end instruments now often use positional information from satellite navigation systems. The rotation and orientation is often provided within a degree or two with electronic compasses. Compasses can measure not just azimuth (i. e. degrees to magnetic north), but also altitude (degrees above the horizon), since the magnetic field curves into the Earth at different angles at different latitudes. More exact orientations require gyroscopic-aided orientation, periodically realigned by different methods including navigation from stars or known benchmarks.

Data Processing

Generally speaking, remote sensing works on the principle of the *inverse problem*. While the object or phenomenon of interest (the state) may not be directly measured, there exists some other variable that can be detected and measured (the observation) which may be related to the object of interest through a calculation. The common analogy given to describe this is trying to determine the type of animal from its footprints. For example, while it is impossible to directly measure temperatures in the upper atmosphere, it is possible to measure the spectral emissions from a known chemical species (such as carbon dioxide) in that region. The frequency of the emissions may then be related via thermodynamics to the temperature in that region.

The quality of remote sensing data consists of its spatial, spectral, radiometric and temporal resolutions.

Spatial Resolution

The size of a pixel that is recorded in a raster image – typically pixels may correspond to square areas ranging in side length from 1 to 1,000 metres (3.3 to 3,280.8 ft).

Spectral Resolution

The wavelength of the different frequency bands recorded – usually, this is related to the number of frequency bands recorded by the platform. Current Landsat collection is that of seven bands, including several in the infrared spectrum, ranging from a spectral resolution of 0.7 to 2.1 µm. The Hyperion sensor on Earth Observing-1 resolves 220 bands from 0.4 to 2.5 µm, with a spectral resolution of 0.10 to 0.11 µm per band.

Radiometric Resolution

The number of different intensities of radiation the sensor is able to distinguish. Typically, this ranges from 8 to 14 bits, corresponding to 256 levels of the gray scale and up to 16,384 intensities or "shades" of colour, in each band. It also depends on the instrument noise.

Temporal Resolution

The frequency of flyovers by the satellite or plane, and is only relevant in time-series studies or those requiring an averaged or mosaic image as in deforesting monitoring. This was first used by the intelligence community where repeated coverage revealed changes in infrastructure, the deployment of units or the modification/introduction of equipment. Cloud cover over a given area or object makes it necessary to repeat the collection of said location.

In order to create sensor-based maps, most remote sensing systems expect to extrapolate sensor data in relation to a reference point including distances between known points on the ground. This depends on the type of sensor used. For example, in conventional photographs, distances are accurate in the center of the image, with the distortion of measurements increasing the farther you get from the center. Another factor is that of the platen against which the film is pressed can cause severe errors when photographs are used to measure ground distances. The step in which this problem is resolved is called georeferencing, and involves computer-aided matching of points in the image (typically 30 or more points per image) which is extrapolated with the use of an established benchmark, "warping" the image to produce accurate spatial data. As of the early 1990s, most satellite images are sold fully georeferenced.

In addition, images may need to be radiometrically and atmospherically corrected.

Radiometric Correction

Allows avoidance of radiometric errors and distortions. The illumination of objects on the Earth surface is uneven because of different properties of the relief. This factor is taken into account in the method of radiometric distortion correction. Radiometric correction gives a scale to the pixel values, e. g. the monochromatic scale of 0 to 255 will be converted to actual radiance values.

Topographic Correction (also Called Terrain Correction)

In rugged mountains, as a result of terrain, the effective illumination of pixels varies considerably. In a remote sensing image, the pixel on the shady slope receives weak illumination and has a low radiance value, in contrast, the pixel on the sunny slope receives strong illumination and has a high radiance value. For the same object, the pixel radiance value on the shady slope will be different from that on the sunny slope. Additionally, different objects may have similar radiance values. These ambiguities seriously affected remote sensing image information extraction accuracy in mountainous areas. It became the main obstacle to further application of remote sensing images. The purpose of topographic correction is to eliminate this effect, recovering the true reflectivity or radiance of objects in horizontal conditions. It is the premise of quantitative remote sensing application.

Atmospheric Correction

Elimination of atmospheric haze by rescaling each frequency band so that its minimum value (usually realised in water bodies) corresponds to a pixel value of 0. The digitizing of data also makes it possible to manipulate the data by changing gray-scale values.

Interpretation is the critical process of making sense of the data. The first application was that of aerial photographic collection which used the following process; spatial measurement through the use of a light table in both conventional single or stereographic coverage, added skills such as the use of photogrammetry, the use of photomosaics, repeat coverage, Making use of objects' known dimensions in order to detect modifications. Image Analysis is the recently developed automated computer-aided application which is in increasing use.

Object-Based Image Analysis (OBIA) is a sub-discipline of GIScience devoted to partitioning remote sensing (RS) imagery into meaningful image-objects, and assessing their characteristics through spatial, spectral and temporal scale.

Old data from remote sensing is often valuable because it may provide the only long-term data for a large extent of geography. At the same time, the data is often complex to interpret, and bulky to store. Modern systems tend to store the data digitally, often with lossless compression. The difficulty with this approach is that the data is fragile, the format may be archaic, and the data may be easy to falsify. One of the best systems for archiving data series is as computer-generated machine-readable ultrafiche, usually in typefonts such as OCR-B, or as digitized half-tone images. Ultrafiches survive well in standard libraries, with lifetimes of several centuries. They can be created, copied, filed and retrieved by automated systems. They are about as compact as archival magnetic media, and yet can be read by human beings with minimal, standardized equipment.

Data Processing Levels

To facilitate the discussion of data processing in practice, several processing "levels" were first defined in 1986 by NASA as part of its Earth Observing System and steadily adopted since then, both internally at NASA (e. g.,) and elsewhere (e. g.,); these definitions are:

Level	Description
0	Reconstructed, unprocessed instrument and payload data at full resolution, with any and all communications artifacts (e. g., synchronization frames, communications headers, duplicate data) removed.
1a	Reconstructed, unprocessed instrument data at full resolution, time-referenced, and annotated with ancillary information, including radiometric and geometric calibration coefficients and georeferencing parameters (e. g., platform ephemeris) computed and appended but not applied to the Level 0 data (or if applied, in a manner that level 0 is fully recoverable from level 1a data).
1b	Level 1a data that have been processed to sensor units (e. g., radar backscatter cross section, brightness temperature, etc.); not all instruments have Level 1b data; level 0 data is not recoverable from level 1b data.
2	Derived geophysical variables (e. g., ocean wave height, soil moisture, ice concentration) at the same resolution and location as Level 1 source data.
3	Variables mapped on uniform spacetime grid scales, usually with some completeness and consistency (e. g., missing points interpolated, complete regions mosaicked together from multiple orbits, etc.).
4	Model output or results from analyses of lower level data (i. e., variables that were not measured by the instruments but instead are derived from these measurements).

A Level 1 data record is the most fundamental (i. e., highest reversible level) data record that has significant scientific utility, and is the foundation upon which all subsequent data sets are produced. Level 2 is the first level that is directly usable for most scientific applications; its value is much greater than the lower levels. Level 2 data sets tend to be less voluminous than Level 1 data because they have been reduced temporally, spatially, or spectrally. Level 3 data sets are generally smaller than lower level data sets and thus can be dealt with without incurring a great deal of data handling overhead. These data tend to be generally more useful for many applications. The regular spatial and temporal organization of Level 3 datasets makes it feasible to readily combine data from different sources.

While these processing levels are particularly suitable for typical satellite data processing pipelines, other data level vocabularies have been defined and may be appropriate for more heterogeneous workflows.

Maps

A map is a visual representation of an entire area or a part of an area, typically represented on a flat surface. The work of a map is to illustrate specific and detailed features of a particular area, most frequently used to illustrate geography. There are many kinds of maps; static, two-dimensional, three-dimensional, dynamic and even interactive. Maps attempt to represent various things, like political boundaries, physical features, roads, topography, population, climates, natural resources and economic activities.

Cartography

The study and practice of the many facets of maps and map making is called Cartography. It can be described as the art and science of map making. Apart from designing and producing maps, cartography includes studying the history of maps, printing, distributing and selling them, collecting, conserving and curating them in map libraries. The variety of maps available goes well beyond road and topographic maps, it incorporates military charts, statistical , geological, tourist and travel maps, weather and climate maps, general and specialist atlases, cartograms, transport network diagrams etc Map for computer and internet use have recently grown in importance and Geographical Information Systems (GIS) have a digital map at their core. Good Cartography is important because a well designed map communicates better than a badly designed one. The quality of each map varies widely.

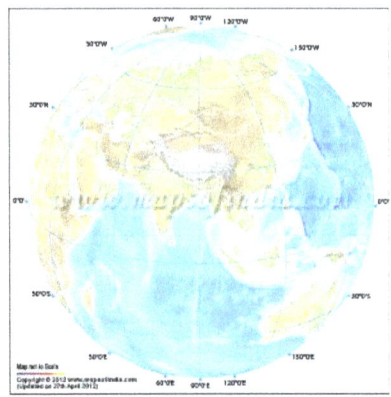

Map and Globe

The word "Globe" is taken up from the Latin word "Globus" which means a round mass or sphere. A globe is a three dimensional representation of the earth that does not distort the shape or size of earth's geographical features. Sometimes a globe shows the topography elevations otherwise they are plain round, some globes include imprinted lines with parallels and meridians for finding the coordinates of a particular place. Globes offer the most precise view of Earth at present.

Advantages of Maps Over Globes

- Maps are more compact
- Easy to store
- Can hold different range of scales
- Easily viewed on different sized screens
- Can show larger portions of an area at once
- Cheaper to produce and transport

Uses of Map

A map is useful for both a layman and an intelligent person, as maps contain loads of information. It is up to an individual how he makes use of it. Maps are generally used for:

- Analysis
- Confirmation
- Communication
- Decoration
- Collection
- Investment
- Exploration
- Hypothesis Stimulation
- Navigation
- Control & Planning
- Map Reading
- Storage of Information
- Historical perspective

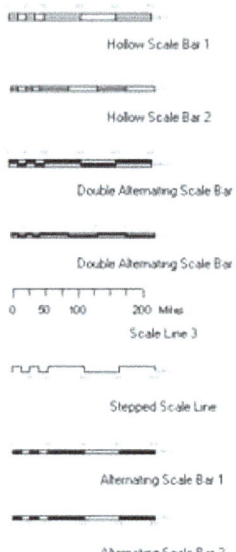

Hollow Scale Bar 1

Hollow Scale Bar 2

Double Alternating Scale Bar 1

Double Alternating Scale Bar 2

0   50   100   200 Miles

Scale Line 3

Stepped Scale Line

Alternating Scale Bar 1

Alternating Scale Bar 2

Scale of Map

As a map has to represent a portion of earth's surface accurately, each map has to have a "scale" which indicates the relation between the distance on the map and the actual distance on the land. The map scale is typically shown in the legend box of a map, along with other symbols that provides useful information about the map.

A ratio or representative fraction specifies how many units on land is equal tone unit on the map. For example, a map showing a scale of 1/ 100000 or 1:100000, tells us that one centimeter on the map is equal to 100000 centimeters i.e. 1 kilometer on the Land. The ratio is always mentioned in the map, such as "one centimeter equals one kilometer" or "one inch equals ten miles".

A graphic scale also known as the bar scale is a line that specifies the distance in kilometers or miles as they show on a map; even as the map is enlarged or reduced in size the line has an advantage of remaining accurate. Maps can be recognized as large scale or small scale maps. A large scale map shows much more detail than a small scale map.

Map Projection and Different Types of Projections

The method of representing the surface of a globe or any three dimensional body on a flat surface is known as projection. Map projection is important for creating maps. The basic problem in any map projection is that there is always some distortion. It could be in distance, shape, area or direction. Map projections can be constructing to conserve one or more of these properties that are area, direction, shape, scale, distance and bearing, but not all of them together. Every projection gives and takes basic metric properties in dissimilar ways.

Projections of Maps are Classified Into Two Types

1) Projections by surface:

- Cylindrical
- Pseudo cylindrical
- Hybrid
- Conical
- Pseudo conical
- Azimuthal

2) Projections by conservation of a metric property:

- Conformal
- Equal-area
- Equidistant
- Gnomonic
- Retro Azimuthal
- Compromise projections.

Topographic Map

A topographic map is a detailed and accurate two-dimensional representation of natural and human-made features on the Earth's surface. These maps are used for a number of applications, from camping, hunting, fishing, and hiking to urban planning, resource management, and surveying. The most distinctive characteristic of a topographic map is that the three-dimensional shape of the Earth's surface is modeled by the use of contour lines. Contours are imaginary lines that connect locations of similar elevation. Contours make it possible to represent the height of mountains and steepness of slopes on a two-dimensional map surface. Topographic maps also use a variety of symbols to describe both natural and human made features such as roads, buildings, quarries, lakes, streams, and vegetation.

Topographic maps produced by the Canadian National Topographic System (NTS) are generally available in two different scales: 1:50,000 and 1:250,000. Maps with a scale of 1:50,000 are relatively large-scale covering an area approximately 1000 square kilometers. At this scale, features as small as a single home can be shown. The smaller scale 1:250,000 topographic map is more of a general purpose reconnaissance-type map. A map of this scale covers the same area of land as sixteen 1:50,000 scale maps.

In the United States, topographic maps have been made by the United States Geological Survey (USGS) since 1879. Topographic coverage of the United States is available at scales of 1:24,000, 1:25,000 (metric), 1:62,250, 1:63,360 (Alaska only), 1:100,000 and 1:250,000.

Topographic Map Symbols

Topographic maps use symbols to represent natural and human constructed features found in the environment. The symbols used to represent features can be of three types: points, lines, and polygons. Points are used to depict features like bridges and buildings. Lines are used to graphically illustrate features that are linear. Some common linear features include roads, railways, and rivers. However, we also need to include representations of area, in the case of forested land or cleared land; this is done through the use of color.

Contour Lines

Topographic maps can describe vertical information through the use of contour lines (contours). A contour line is an isoline that connects points on a map that have the same elevation. Contours are often drawn on a map at a uniform vertical distance. This distance is called the contour interval. The map in the Figure shows contour lines with an interval of 100 feet. Note that every fifth brown contour lines is drawn bold and has the appropriate elevation labeled on it. These contours are called index contours. On Figure they represent elevations of 500, 1000, 1500, 2000 feet and so on. The interval at which contours are drawn on a map depends on the amount of the relief depicted and the scale of the map.

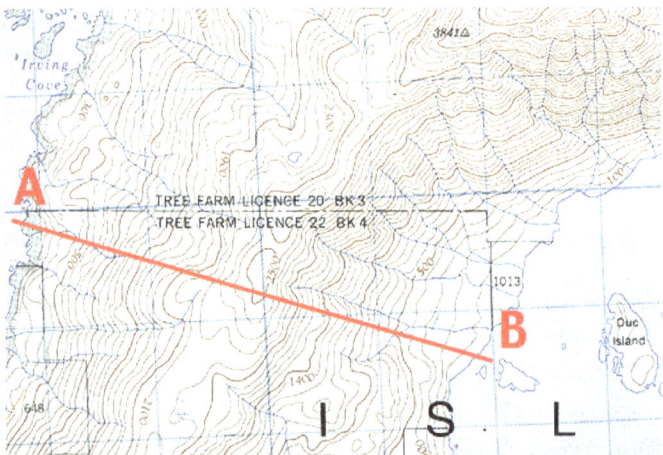

Figure: Portion of the "Tofino" 1:50,000 National Topographic Series of Canada map.

The brown lines drawn on this map are contour lines. Each line represents a vertical increase in elevation of 100 feet. The bold brown contour lines are called index contours. The index contours are labeled with their appropriate elevation which increases at a rate of 500 feet. Note the blue line drawn to separate water from land represents an elevation of 0 feet or sea-level.

Contour lines provide us with a simple effective system for describing landscape configuration on a two-dimensional map. The arrangement, spacing, and shape of the contours provide the user of the map with some idea of what the actual topographic configuration of the land surface looks like. Contour intervals the are spaced closely together describe a steep slope. Gentle slopes are indicated by widely spaced contours. Contour lines that V upwards indicate the presence of a river valley. Ridges are shown by contours that V downwards.

Topographic Profiles

A topographic profile is a two-dimensional diagram that describes the landscape in vertical cross-section. Topographic profiles are often created from the contour information found on topographic maps. The simplest way to construct a topographic profile is to place a sheet of blank paper along a horizontal transect of interest. From the map, the elevation of the various contours is transferred on to the edge of the paper from one end of the transect to the other. Now on a sheet of graph paper use the x-axis to represent the horizontal distance covered by the transect. The y-axis is used to represent the vertical dimension and measures the change in elevation along the transect. Most people exaggerate the measure of elevation on the y-axis to make changes in relief stand out. Place the beginning of the transect as copied on the piece of paper at the intersect of the x and y-axis on the graph paper. The contour information on the paper's edge is now copied onto the piece of graph paper. Figure shows a topographic profile drawn from the information found on the transect A-B above.

Figure: The following topographic profile shows the vertical change in surface elevation along the transect AB from figure. A vertical exaggeration of about 4.2 times was used in the profile (horizontal scale = 1:50,000, vertical scale = 1:12,000 and vertical exaggeration = horizontal scale/vertical scale).

Military Grid Reference System and Map Location

Two rectangular grid systems are available on topographic maps for identifying the location of points. These systems are the Universal Transverse Mercator (UTM) grid system and the Military Grid Reference System. The Military Grid Reference System is a simplified form of Universal Transverse Mercator grid system and it provides a very quick and easy method of referencing a location on a topographic map. On a topographic maps with a scale 1:50,000 and larger, the Military Grid Reference System is superimposed on the surface of map as blue colored series of equally spaced horizontal and vertical lines. Identifying numbers for each of these lines is found along the map's margin. Each identifying number consists of two digits which range from a value of 00 to 99 (Figure below). Each individual square in the grid system represents a distance of a 1000 by 1000 meters and the total size of the grid is 100,000 by 100,000 meters.

One problem associated with the Military Grid Reference System is the fact that reference numbers must be repeated every 100,000 meters. To overcome this difficulty, a method was devised to identify each 100,000 by 100,000 meter grid with two identifying letters which are printed in blue on the border of all topographic maps (note some maps may show more than one grid). When making reference to a location with the Military Grid Reference System identifying letters are always given before the horizontal and vertical coordinate numbers.

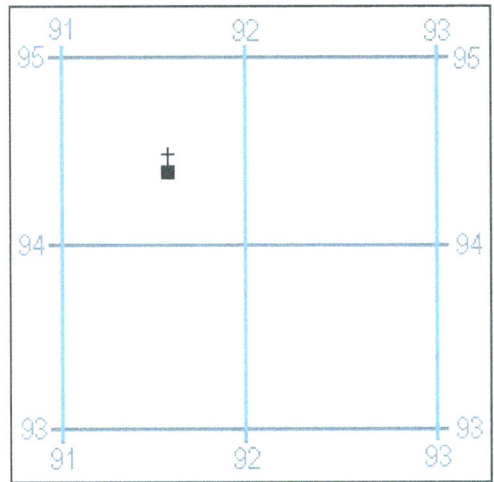

Figure: Portion of a Military Grid Reference System found on a topographic map.

Coordinates on this system are based on a X (horizontal increasing from left to right) and Y (vertical increasing from bottom to top) system. The symbol depicting a church is located in the square 9194. Note that the value along the X-axis (easting) is given first followed by the value on the Y-axis (northing).

Each individual square in the Military Grid Reference System can be further divided into 100 smaller squares (ten by ten). This division allows us to calculate the location of an object to within 100 meters. Figure indicates that the church is six tenths of the way between lines 91 and 92, and four tenths of the way between lines 94 and 95. Using these values, we can state that the easting as being 916 and the northing as 944. By convention, these two numbers are combined into a coordinate reference of 916944.

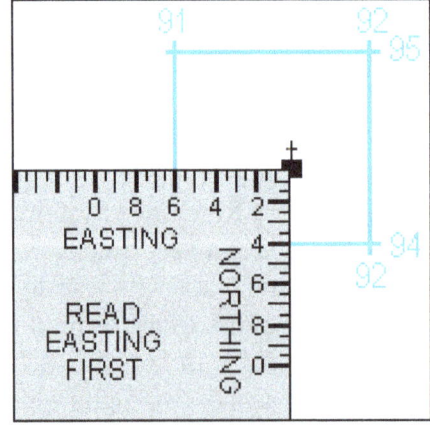

Figure: Further determination of the location of the church described in figure. Using the calibrated ruler we can now suggest the location of the church to be 916 on the X-axis and 944 on the Y-axis.

GIS

A geographic information system (GIS) is a system designed to capture, store, manipulate, analyze, manage, and present all types of geographical data. The key word to this technology is Geography – this means that some portion of the data is spatial. In other words, data that is in some way referenced to locations on the earth.

Coupled with this data is usually tabular data known as attribute data. Attribute data can be generally defined as additional information about each of the spatial features. An example of this would be schools. The actual location of the schools is the spatial data. Additional data such as the school name, level of education taught, student capacity would make up the attribute data.

It is the partnership of these two data types that enables GIS to be such an effective problem solving tool through spatial analysis.

GIS is more than just software. People and methods are combined with geospatial software and tools, to enable spatial analysis, manage large datasets, and display information in a map/graphical form.

GIS can be used as tool in both problem solving and decision making processes, as well as for visualization of data in a spatial environment. Geospatial data can be analyzed to determine (1) the location of features and relationships to other features, (2) where the most and/or least of some feature exists, (3) the density of features in a given space, (4) what is happening inside an area of interest (AOI), (5) what is happening nearby some feature or phenomenon, and (6) and how a specific area has changed over time (and in what way).

1.   Mapping where things are. We can map the spatial location of real-world features and visualize the spatial relationships among them. Example: below we see a map of frac sand mine locations and sandstone areas in Wisconsin. We can see visual patterns in the data by determining that frac sand mining activity occurs in a region with a specific type of geology.

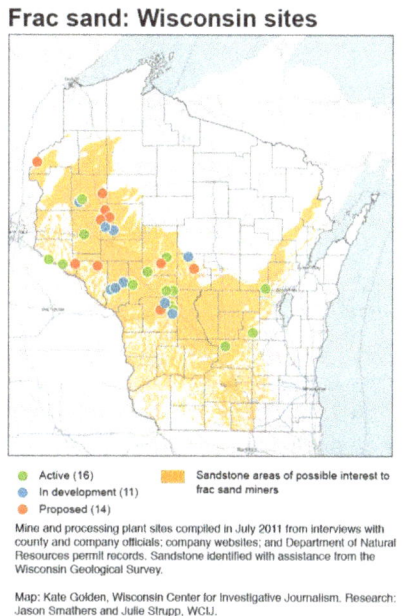

Frac sand: Wisconsin sites

● Active (16)   Sandstone areas of possible interest to
● In development (11)   frac sand miners
● Proposed (14)

Mine and processing plant sites compiled in July 2011 from interviews with
county and company officials; company websites; and Department of Natural
Resources permit records. Sandstone identified with assistance from the
Wisconsin Geological Survey.

Map: Kate Golden, Wisconsin Center for Investigative Journalism. Research:
Jason Smathers and Julie Strupp, WCIJ.

2. Mapping quantities. People map quantities, such as where the most and least are, to find places that meet their criteria or to see the relationships between places.

   Example: below is a map of cemetery locations in Wisconsin. The map shows the cemetery locations as dots (dot density) and each county is color coded to show where the most and least are (lighter blue means fewer cemeteries).

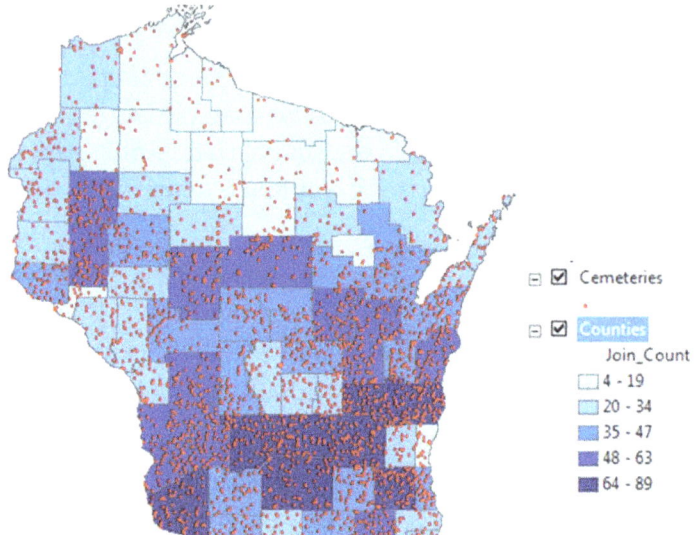

3. Mapping densities. Sometimes it is more important to map concentrations, or a quantity normalized by area or total number. Example: Below we have mapped the population density of Manhattan (total population counts normalized by the area in sq. miles of census tracts.)

4. Finding what is inside. We can use GIS to determine what is happening or what features are located inside a specific area/region. We can determine the characteristics of "inside" by creating specific criteria to define an area of interest (AOI). Example: below is a map showing a flood event and the tax parcels and buildings in the floodway. We can use tools like CLIP to determine which parcels fall inside the flood event. Further, we can use attributes of the parcels to determine potential costs of property damage.

5. Finding what is nearby. We can find out what is happening within a set distance of a feature or event by mapping what is nearby using geoprocessing tools like BUFFER. Example: below we see a map of drive times from a central location in the City of Madison, WI. We can use streets as a network and add specific criteria like speed limit and intersection controls to determine how far a driver can typically get in 5, 10, or 15 minutes. (Map courtesy of UW Extension)

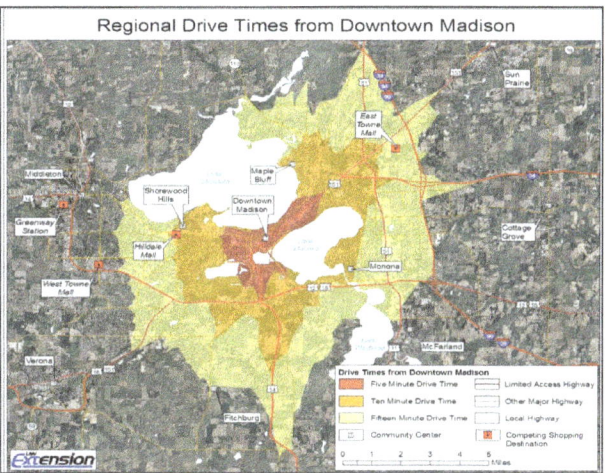

6. Mapping change. We can map the change in a specific geographic area to anticipate future conditions, decide on a course of action, or to evaluate the results of an action or policy. Example: below we see land use maps of Barnstable, MA showing changes in residential development from 1951 to 1999. The dark green shows forest, while bright yellow shows residential development. Applications like this can help inform community planning processes and policies.

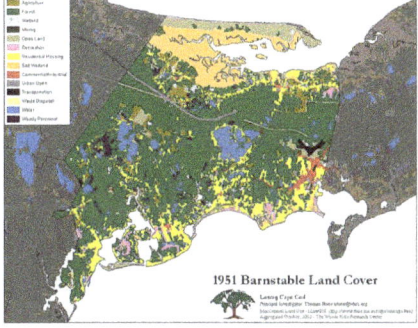

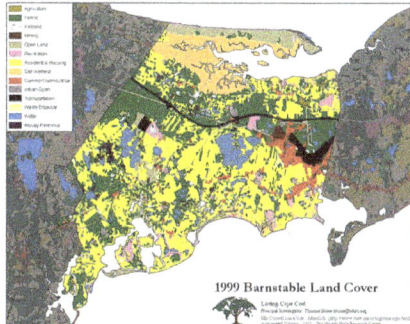

Techniques and Technology

Modern GIS technologies use digital information, for which various digitized data creation methods are used. The most common method of data creation is digitization, where a hard copy map or survey plan is transferred into a digital medium through the use of a CAD program, and geo-referencing capabilities. With the wide availability of ortho-rectified imagery (from satellites, aircraft, Helikites and UAVs), heads-up digitizing is becoming the main avenue through which geographic data is extracted. Heads-up digitizing involves the tracing of geographic data directly on top of the aerial imagery instead of by the traditional method of tracing the geographic form on a separate digitizing tablet (heads-down digitizing).

Relating Information from Different Sources

GIS uses spatio-temporal (space-time) location as the key index variable for all other information. Just as a relational database containing text or numbers can relate many different tables using common key index variables, GIS can relate otherwise unrelated information by using location as the key index variable. The key is the location and/or extent in space-time.

Any variable that can be located spatially, and increasingly also temporally, can be referenced using a GIS. Locations or extents in Earth space–time may be recorded as dates/times of occurrence, and x, y, and z coordinates representing, longitude, latitude, and elevation, respectively. These GIS coordinates may represent other quantified systems of temporo-spatial reference (for example, film frame number, stream gage station, highway mile-marker, surveyor benchmark, building address, street intersection, entrance gate, water depth sounding, POS or CAD drawing origin/units). Units applied to recorded temporal-spatial data can vary widely, but all Earth-based spatial–temporal location and extent references should, ideally, be relatable to one another and ultimately to a "real" physical location or extent in space–time.

Related by accurate spatial information, an incredible variety of real-world and projected past or future data can be analyzed, interpreted and represented. This key characteristic of GIS has begun to open new avenues of scientific inquiry into behaviors and patterns of real-world information that previously had not been systematically correlated.

GIS Uncertainties

GIS accuracy depends upon source data, and how it is encoded to be data referenced. Land surveyors have been able to provide a high level of positional accuracy utilizing the GPS-derived positions. High-resolution digital terrain and aerial imagery, powerful computers and Web technology are changing the quality, utility, and expectations of GIS to serve society on a grand scale, but nevertheless there are other source data that affect overall GIS accuracy like paper maps, though these may be of limited use in achieving the desired accuracy.

In developing a digital topographic database for a GIS, topographical maps are the main source, and aerial photography and satellite imagery are extra sources for collecting data and identifying attributes which can be mapped in layers over a location facsimile of scale. The scale of a map and geographical rendering area representation type are very important aspects since the information content depends mainly on the scale set and resulting locatability of the map's representations. In order to digitize a map, the map has to be checked within theoretical dimensions, then scanned into a raster format, and resulting raster data has to be given a theoretical dimension by a rubber sheeting/warping technology process.

A quantitative analysis of maps brings accuracy issues into focus. The electronic and other equipment used to make measurements for GIS is far more precise than the machines of conventional map analysis. All geographical data are inherently inaccurate, and these inaccuracies will propagate through GIS operations in ways that are difficult to predict.

Data Representation

GIS data represents real objects (such as roads, land use, elevation, trees, waterways, etc.) with digital data determining the mix. Real objects can be divided into two abstractions: discrete objects (e.g., a house) and continuous fields (such as rainfall amount, or elevations). Traditionally, there are two broad methods used to store data in a GIS for both kinds of abstractions mapping references: raster images and vector. Points, lines, and polygons are the stuff of mapped location attribute references. A new hybrid method of storing data is that of identifying point clouds, which combine three-dimensional points with RGB information at each point, returning a "3D color image". GIS thematic maps then are becoming more and more realistically visually descriptive of what they set out to show or determine.

Data Capture

Data capture—entering information into the system—consumes much of the time of GIS practitioners. There are a variety of methods used to enter data into a GIS where it is stored in a digital format.

Existing data printed on paper or PET film maps can be digitized or scanned to produce digital data. A digitizer produces vector data as an operator traces points, lines, and polygon boundaries from a map. Scanning a map results in raster data that could be further processed to produce vector data.

Survey data can be directly entered into a GIS from digital data collection systems on survey instruments using a technique called coordinate geometry (COGO). Positions from a global navigation satellite system (GNSS) like Global Positioning System can also be collected and then imported into a GIS. A current trend in data collection gives users the ability to utilize field computers with the ability to edit live data using wireless connections or disconnected editing sessions. This has been enhanced by the availability of low-cost mapping-grade GPS units with decimeter accuracy in real time. This eliminates the need to post process, import, and update the data in the office after fieldwork has been collected. This includes the ability to incorporate positions collected using a laser rangefinder. New technologies also allow users to create maps as well as analysis directly in the field, making projects more efficient and mapping more accurate.

Remotely sensed data also plays an important role in data collection and consist of sensors attached to a platform. Sensors include cameras, digital scanners and lidar, while platforms usually consist of aircraft and satellites. In England in the mid 1990s, hybrid kite/balloons called helikites first pioneered the use of compact airborne digital cameras as airborne geo-information systems. Aircraft measurement software, accurate to 0.4 mm was used to link the photographs and measure the ground. Helikites are inexpensive and gather more accurate data than aircraft. Helikites can be used over roads, railways and towns where unmanned aerial vehicles (UAVs) are banned.

Recently aerial data collection is becoming possible with miniature UAVs. For example, the Aery-on Scout was used to map a 50-acre area with a ground sample distance of 1 inch (2.54 cm) in only 12 minutes.

The majority of digital data currently comes from photo interpretation of aerial photographs. Soft-copy workstations are used to digitize features directly from stereo pairs of digital photographs. These systems allow data to be captured in two and three dimensions, with elevations measured directly from a stereo pair using principles of photogrammetry. Analog aerial photos must be scanned before being entered into a soft-copy system, for high-quality digital cameras this step is skipped.

Satellite remote sensing provides another important source of spatial data. Here satellites use different sensor packages to passively measure the reflectance from parts of the electromagnetic spectrum or radio waves that were sent out from an active sensor such as radar. Remote sensing collects raster data that can be further processed using different bands to identify objects and classes of interest, such as land cover.

When data is captured, the user should consider if the data should be captured with either a relative accuracy or absolute accuracy, since this could not only influence how information will be interpreted but also the cost of data capture.

After entering data into a GIS, the data usually requires editing, to remove errors, or further processing. For vector data it must be made "topologically correct" before it can be used for some advanced analysis. For example, in a road network, lines must connect with nodes at an intersection. Errors such as undershoots and overshoots must also be removed. For scanned maps, blemishes on the source map may need to be removed from the resulting raster. For example, a fleck of dirt might connect two lines that should not be connected.

Example of hardware for mapping (GPS and laser rangefinder) and data collection. The current trend for geographical information system (GIS) is that accurate mapping and data analysis are completed while in the field. Depicted hardware (field-map technology) is used mainly for forest inventories, monitoring and mapping.

Raster-to-vector Translation

Data restructuring can be performed by a GIS to convert data into different formats. For example, a GIS may be used to convert a satellite image map to a vector structure by generating lines around all cells with the same classification, while determining the cell spatial relationships, such as adjacency or inclusion.

More advanced data processing can occur with image processing, a technique developed in the late 1960s by NASA and the private sector to provide contrast enhancement, false color rendering and a variety of other techniques including use of two dimensional Fourier transforms. Since digital data is collected and stored in various ways, the two data sources may not be entirely compatible. So a GIS must be able to convert geographic data from one structure to another. In so doing, the implicit assumptions behind different ontologies and classifications require analysis. Object ontologies have gained increasing prominence as a consequence of object-oriented programming and sustained work by Barry Smith and co-workers.

Projections, Coordinate Systems, and Registration

The earth can be represented by various models, each of which may provide a different set of coordinates (e.g., latitude, longitude, elevation) for any given point on the Earth's surface. The simplest model is to assume the earth is a perfect sphere. As more measurements of the earth have accumulated, the models of the earth have become more sophisticated and more accurate. In fact, there are models called datums that apply to different areas of the earth to provide increased accuracy, like NAD83 for U.S. measurements, and the World Geodetic System for worldwide measurements.

Spatial Analysis with Geographical Information System (GIS)

GIS spatial analysis is a rapidly changing field, and GIS packages are increasingly including analytical tools as standard built-in facilities, as optional toolsets, as add-ins or 'analysts'. In many instances these are provided by the original software suppliers (commercial vendors or collaborative non commercial development teams), while in other cases facilities have been developed and are provided by third parties. Furthermore, many products offer software development kits (SDKs), programming languages and language support, scripting facilities and/or special interfaces for developing one's own analytical tools or variants. The website "Geospatial Analysis" and associated book/ebook attempt to provide a reasonably comprehensive guide to the subject. The increased availability has created a new dimension to business intelligence termed "spatial intelligence" which, when openly delivered via intranet, democratizes access to geographic and social network data. Geospatial intelligence, based on GIS spatial analysis, has also become a key element for security. GIS as a whole can be described as conversion to a vectorial representation or to any other digitisation process.

Slope and Aspect

Slope can be defined as the steepness or gradient of a unit of terrain, usually measured as an angle in degrees or as a percentage. Aspect can be defined as the direction in which a unit of terrain faces. Aspect is usually expressed in degrees from north. Slope, aspect, and surface curvature in terrain analysis are all derived from neighborhood operations using elevation values of a cell's adjacent neighbours. Slope is a function of resolution, and the spatial resolution used to calculate slope and aspect should always be specified. Various authors have compared techniques for calculating slope and aspect.

The following method can be used to derive slope and aspect: The elevation at a point or unit of terrain will have perpendicular tangents (slope) passing through the point, in an east-west and north-south direction. These two tangents give two components, $\partial z/\partial x$ and $\partial z/\partial y$, which then be used to determine the overall direction of slope, and the aspect of the slope. The gradient is defined as a vector quantity with components equal to the partial derivatives of the surface in the x and y directions.

The calculation of the overall 3x3 grid slope S and aspect A for methods that determine east-west and north-south component use the following formulas respectively:

$$\tan S = \sqrt{\left(\frac{\partial z}{\partial x}\right)^2 + \left(\frac{\partial z}{\partial y}\right)^2}$$

$$A = \left|\frac{\left(\frac{-\partial z}{\partial y}\right)}{\left(\frac{\partial z}{\partial x}\right)}\right|$$

Zhou and Liu describe another formula for calculating aspect, as follows:

$$A = 270° + \arctan\left(\frac{\left(\frac{\partial z}{\partial x}\right)}{\left(\frac{\partial z}{\partial y}\right)}\right) - 90°\left(\frac{\left(\frac{\partial z}{\partial y}\right)}{\left|\frac{\partial z}{\partial y}\right|}\right)$$

Data Analysis

It is difficult to relate wetlands maps to rainfall amounts recorded at different points such as airports, television stations, and schools. A GIS, however, can be used to depict two- and three-dimensional characteristics of the Earth's surface, subsurface, and atmosphere from information points. For example, a GIS can quickly generate a map with isopleth or contour lines that indicate differing amounts of rainfall. Such a map can be thought of as a rainfall contour map. Many sophisticated methods can estimate the characteristics of surfaces from a limited number of point measurements. A two-dimensional contour map created from the surface modeling of rainfall point measurements may be overlaid and analyzed with any other map in a GIS covering the same area. This GIS derived map can then provide additional information - such as the viability of water power potential as a renewable energy source. Similarly, GIS can be used to compare other renewable energy resources to find the best geographic potential for a region.

Additionally, from a series of three-dimensional points, or digital elevation model, isopleth lines representing elevation contours can be generated, along with slope analysis, shaded relief, and other elevation products. Watersheds can be easily defined for any given reach, by computing all of the areas contiguous and uphill from any given point of interest. Similarly, an expected thalweg of where surface water would want to travel in intermittent and permanent streams can be computed from elevation data in the GIS.

Topological Modeling

A GIS can recognize and analyze the spatial relationships that exist within digitally stored spatial data. These topological relationships allow complex spatial modelling and analysis to be performed. Topological relationships between geometric entities traditionally include adjacency (what adjoins what), containment (what encloses what), and proximity (how close something is to something else).

Geometric Networks

Geometric networks are linear networks of objects that can be used to represent interconnected features, and to perform special spatial analysis on them. A geometric network is composed of edges, which are connected at junction points, similar to graphs in mathematics and computer science. Just like graphs, networks can have weight and flow assigned to its edges, which can be used to represent various interconnected features more accurately. Geometric networks are often used to model road networks and public utility networks, such as electric, gas, and water networks. Network modeling is also commonly employed in transportation planning, hydrology modeling, and infrastructure modeling.

Hydrological Modeling

GIS hydrological models can provide a spatial element that other hydrological models lack, with the analysis of variables such as slope, aspect and watershed or catchment area. Terrain analysis is fundamental to hydrology, since water always flows down a slope. As basic terrain analysis of a digital elevation model (DEM) involves calculation of slope and aspect, DEMs are very useful for hydrological analysis. Slope and aspect can then be used to determine direction of surface runoff, and hence flow accumulation for the formation of streams, rivers and lakes. Areas of divergent flow can also give a clear indication of the boundaries of a catchment. Once a flow direction and accumulation matrix has been created, queries can be performed that show contributing or dispersal areas at a certain point. More detail can be added to the model, such as terrain roughness, vegetation types and soil types, which can influence infiltration and evapotranspiration rates, and hence influencing surface flow. One of the main uses of hydrological modeling is in environmental contamination research. Other applications of hydrological modeling include groundwater and surface water mapping, as well as flood risk maps.

Cartographic Modeling

Dana Tomlin probably coined the term "cartographic modeling" in his PhD dissertation (1983); he later used it in the title of his book, *Geographic Information Systems and Cartographic Modeling* (1990). Cartographic modeling refers to a process where several thematic layers of the same area are produced, processed, and analyzed. Tomlin used raster layers, but the overlay method can be used more generally. Operations on map layers can be combined into algorithms, and eventually into simulation or optimization models.

An example of use of layers in a GIS application.

In this example, the forest-cover layer (light green) forms the bottom layer, with the topographic layer (contour lines) over it. Next up is a standing water layer (pond, lake) and then a flowing water layer (stream, river), followed by the boundary layer and finally the road layer on top. The order is very important in order to properly display the final result. Note that the ponds are layered under the streams, so that a stream line can be seen overlying one of the ponds.

Map Overlay

The combination of several spatial datasets (points, lines, or polygons) creates a new output vector dataset, visually similar to stacking several maps of the same region. These overlays are similar to mathematical Venn diagram overlays. A union overlay combines the geographic features and attribute tables of both inputs into a single new output. An intersect overlay defines the area where both inputs overlap and retains a set of attribute fields for each. A symmetric difference overlay defines an output area that includes the total area of both inputs except for the overlapping area.

Data extraction is a GIS process similar to vector overlay, though it can be used in either vector or raster data analysis. Rather than combining the properties and features of both datasets, data extraction involves using a "clip" or "mask" to extract the features of one data set that fall within the spatial extent of another dataset.

In raster data analysis, the overlay of datasets is accomplished through a process known as "local operation on multiple rasters" or "map algebra", through a function that combines the values of each raster's matrix. This function may weigh some inputs more than others through use of an "index model" that reflects the influence of various factors upon a geographic phenomenon.

Geostatistics

Geostatistics is a branch of statistics that deals with field data, spatial data with a continuous index. It provides methods to model spatial correlation, and predict values at arbitrary locations (interpolation).

When phenomena are measured, the observation methods dictate the accuracy of any subsequent analysis. Due to the nature of the data (e.g. traffic patterns in an urban environment; weather patterns over the Pacific Ocean), a constant or dynamic degree of precision is always lost in the measurement. This loss of precision is determined from the scale and distribution of the data collection.

To determine the statistical relevance of the analysis, an average is determined so that points (gradients) outside of any immediate measurement can be included to determine their predicted behavior. This is due to the limitations of the applied statistic and data collection methods, and interpolation is required to predict the behavior of particles, points, and locations that are not directly measurable.

Interpolation is the process by which a surface is created, usually a raster dataset, through the input of data collected at a number of sample points. There are several forms of interpolation, each which treats the data differently, depending on the properties of the data set. In comparing interpolation methods, the first consideration should be whether or not the source data will change (exact or approximate). Next is whether the method is subjective, a human interpretation, or objective. Then there is the nature of transitions between points: are they abrupt or gradual. Finally, there is whether a method is global (it uses the entire data set to form the model), or local where an algorithm is repeated for a small section of terrain.

Interpolation is a justified measurement because of a spatial auto correlation principle that recognizes that data collected at any position will have a great similarity to, or influence of those locations within its immediate vicinity.

Digital elevation models, triangulated irregular networks, edge-finding algorithms, Thiessen polygons, Fourier analysis, (weighted) moving averages, inverse distance weighting, kriging, spline, and trend surface analysis are all mathematical methods to produce interpolative data.

Hillshade model derived from a Digital Elevation Model of the Valestra area in the northern Apennines (Italy)

Address Geocoding

Geocoding is interpolating spatial locations (X,Y coordinates) from street addresses or any other spatially referenced data such as ZIP Codes, parcel lots and address locations. A reference theme is required to geocode individual addresses, such as a road centerline file with address ranges. The individual address locations have historically been interpolated, or estimated, by examining address ranges along a road segment. These are usually provided in the form of a table or database. The software will then place a dot approximately where that address belongs along the segment of centerline. For example, an address point of 500 will be at the midpoint of a line segment that starts with address 1 and ends with address 1,000. Geocoding can also be applied against actual parcel data, typically from municipal tax maps. In this case, the result of the geocoding will be an actually positioned space as opposed to an interpolated point. This approach is being increasingly used to provide more precise location information.

Reverse Geocoding

Reverse geocoding is the process of returning an estimated street address number as it relates to a given coordinate. For example, a user can click on a road centerline theme (thus providing a coordinate) and have information returned that reflects the estimated house number. This house number is interpolated from a range assigned to that road segment. If the user clicks at the midpoint of a segment that starts with address 1 and ends with 100, the returned value will be somewhere near 50. Note that reverse geocoding does not return actual addresses, only estimates of what should be there based on the predetermined range.

Multi-criteria Decision Analysis

Coupled with GIS, multi-criteria decision analysis methods support decision-makers in analysing a set of alternative spatial solutions, such as the most likely ecological habitat for restoration, against multiple criteria, such as vegetation cover or roads. MCDA uses decision rules to aggregate the criteria, which allows the alternative solutions to be ranked or prioritised. GIS MCDA may reduce costs and time involved in identifying potential restoration sites.

Data Output and Cartography

Cartography is the design and production of maps, or visual representations of spatial data. The vast majority of modern cartography is done with the help of computers, usually using GIS but production of quality cartography is also achieved by importing layers into a design program to refine it. Most GIS software gives the user substantial control over the appearance of the data.

Cartographic work serves two major functions:

- First, it produces graphics on the screen or on paper that convey the results of analysis to the people who make decisions about resources. Wall maps and other graphics can be generated, allowing the viewer to visualize and thereby understand the results of analyses or simulations of potential events. Web Map Servers facilitate distribution of generated maps through web browsers using various implementations of web-based application programming interfaces (AJAX, Java, Flash, etc.).

- Second, other database information can be generated for further analysis or use. An example would be a list of all addresses within one mile (1.6 km) of a toxic spill.

Graphic Display Techniques

Traditional maps are abstractions of the real world, a sampling of important elements portrayed on a sheet of paper with symbols to represent physical objects. People who use maps must interpret these symbols. Topographic maps show the shape of land surface with contour lines or with shaded relief.

Today, graphic display techniques such as shading based on altitude in a GIS can make relationships among map elements visible, heightening one's ability to extract and analyze information. For example, two types of data were combined in a GIS to produce a perspective view of a portion of San Mateo County, California.

- The digital elevation model, consisting of surface elevations recorded on a 30-meter horizontal grid, shows high elevations as white and low elevation as black.

- The accompanying Landsat Thematic Mapper image shows a false-color infrared image looking down at the same area in 30-meter pixels, or picture elements, for the same coordinate points, pixel by pixel, as the elevation information.

A GIS was used to register and combine the two images to render the three-dimensional perspective view looking down the San Andreas Fault, using the Thematic Mapper image pixels, but shaded using the elevation of the landforms. The GIS display depends on the viewing point of the observer and time of day of the display, to properly render the shadows created by the sun's rays at that latitude, longitude, and time of day.

An archeochrome is a new way of displaying spatial data. It is a thematic on a 3D map that is applied to a specific building or a part of a building. It is suited to the visual display of heat-loss data.

Spatial ETL

Spatial ETL tools provide the data processing functionality of traditional extract, transform, load (ETL) software, but with a primary focus on the ability to manage spatial data. They provide GIS users with the ability to translate data between different standards and proprietary formats, whilst geometrically transforming the data en route. These tools can come in the form of add-ins to existing wider-purpose software such as spreadsheets.

GIS Data Mining

GIS or spatial data mining is the application of data mining methods to spatial data. Data mining, which is the partially automated search for hidden patterns in large databases, offers great potential benefits for applied GIS-based decision making. Typical applications include environmental monitoring. A characteristic of such applications is that spatial correlation between data measurements require the use of specialized algorithms for more efficient data analysis.

GPS is a navigation technology which, by use of satellites, tells the precise information about a location. Basically a GPS system consists of group of satellites and well developed tools such as receiver. The system, however, should comprise at least four satellites. Each satellite and the receiver are equipped with stable atomic clock. The satellite clocks are synchronised with each other and ground clocks. GPS receiver also has a clock but it is not synchronized and is not stable (less stable). Any deviation of actual time of satellites from ground clock should be corrected daily. Four unknown quantities (three coordinates and clock deviation from satellite time) are required to be computed from the synchronized network of satellites and the receiver. The work of the GPS receiver is to receive signals from the network of satellites to compute three basic unknown equations of time and position.

A GPS signal includes a pseudorandom codes and time of transmission and satellite position at that time. The signal broadcasted by GPS is also called carrier frequency with modulation. Further, a pseudorandom code is a sequence of zeros and ones. Practically, the receiver position and the offset of receiver clock relative to receiver system time are computed simultaneously, using the navigation equations to process time of flight (TOFs). TOF is the four values that the receiver forms using time of arrival and time of transmission of the signal. The location is usually converted to latitude, longitude and height relative to geoids (essentially, mean sea level). Then the coordinates are displayed on the screen.

Elements of GPS

The structure of the GPS is a complex one. It consists of three major segments of a space segment, a control segment and a user segment. Launching the satellite into medium earth orbit is a strenuous job. The space segment comprises 24 to 32 satellites or space vehicles in the same orbit, 8 each in three circular orbits. At least six satellites are always in line of sight from almost everywhere on earth's surface.

Next to space segment is control segment. In control segment there is a master control station, an alternate master control station, ground antennae and monitor station. The user segment is composed of thousands of civil, commercial and military positioning service. A GPS receiver or devise consists of an antenna, tuned to the frequency transmitted by satellites. It also includes display screen to provide location and time.

A GPS receiver is classified on the number of satellites it can monitor simultaneously, that is number of channels. Receivers generally have four to five channels but recent advancements have shown that up to 20 channels have also been made.

Satellite frequency: All satellite broadcast frequencies. The frequency band comprises five types such as L1, L2, L3, L4, and L5. These bands have frequency ranges between 1176 MHz to 1600 MHz.

Works of GPS

GPS satellites rotate all around the earth two times in a day. It revolves around in a very accurate course and sends out indication and information to the earth. The receivers of GPS get all the information and apply triangulation to discover the accurate location of the user. Fundamentally, the receiver of GPS contrasts the duration at which a signal was spread by a satellite and allots the time it was received. The time difference formulates how far the receiver is away from the satellites of the GPS. It measures the exact distance with few more satellites and the receiver determines the position of the user and displays it on the map of the electronic appliance.

The receiver must be locked to the signal with at least three satellites to produce a two dimensional position and also tracks the movement of the user. By using four or more satellites, the receiver can determine the three dimensional position of the user which consists of altitude, latitude and longitude. After determining the position of user, the GPS unit calculates other information such as speed, bearing, track, distance, destination, sunrise, and sunset time.

Sources of GPS Signal Errors

Factors that can corrupt the precision of GPS signals and thus influence accurateness incorporate the following:

- Ionosphere and troposphere delays - The satellite signal slows down as it crosses the layers of atmosphere. The GPS system uses a built in model that is used to calculate the regular duration of hindrance required to correct this type of inaccuracy.

- Signal multipath- This error is occurred when the signal is reflected from the objects like taller buildings and larger rocks before it reaches the receiver. This increases the overall time duration of the travel of signal and causes errors and inaccuracy.

- Orbital errors – These errors are also known as ephemeris errors which are used to calculate the inaccuracies of the location of the satellite.

- Number of satellites visible- accuracy depends on the exact number of satellites that a GPS receiver can see. The factors like buildings, terrain, electronic interference blocks the signal accuracy and reception which causes error in position and sometimes no reading in signals. It typically does not work indoors, underwater and underground.

Applications

Not only for military use is a GPS machine widely known for its use in civil and commercial services. Some civilian applications are:

1. Astronomy: Used in Astrometry and celestial mechanics calculations.

2. Automated vehicles: It is also used in automated vehicles (driverless vehicles) to apply locations for cars and trucks.

3. Cellular telephony: Modern mobile phones come equipped with GPS tracking software. It is present because one can know one's position and can also track nearby utilities such as ATMs, coffee shops, restraints, etc. The first cell phone enabled GPS was launched in 1990s. In cellular telephony it is also used in detection for emergency calls and many other applications.

4. Disaster relief and other emergency services: In case of any natural disaster, a GPS is a best tool to identify the location. Even prior to the disasters like cyclones, GPS helps in calculating the estimated time.

5. Fleet tracking: GPS is a developer tool known for its potential to track military ships during the war time.

6. Car location: A GPS enabled car makes it easier to track its location.

7. Geo fencing: In geo fencing, we use GPS to track a human, an animal or a car. The devise is attached to the vehicle, person or on animal's collar. It provides continuous tracking and updating.

8. Geo tagging: one of the major applications is geotagging meaning applying local coordinates to digital objects.

9. GPS for mining: Uses centimetre-level positioning accuracy.

10. GPS tours: helps in determining location of nearby point of interests.

11. Surveying: Surveyors make use of Global Positioning System to plot maps.

References

- Schowengerdt, Robert A. (2007). Remote sensing: models and methods for image processing (3rd ed.). Academic Press. p. 2. ISBN 978-0-12-369407-2

- "A survey of landmine detection using hyperspectral imaging". ISPRS Journal of Photogrammetry and Remote Sensing. 124: 40–53. 2017-02-01. doi:10.1016/j.isprsjprs.2016.12.009. ISSN 0924-2716

- What-is-remote-sensing-in-geography: worldatlas.com, Retrieved 10 July 2018

- "Rapport sur la marche et les effets du choléra dans Paris et le département de la Seine. Année 1832". Gallica. Retrieved 10 May 2012

- Marwick, Ben; Hiscock, Peter; Sullivan, Marjorie; Hughes, Philip (July 2017). "Landform boundary effects on Holocene forager landscape use in arid South Australia". Journal of Archaeological Science: Reports. doi:10.1016/j.jasrep.2017.07.004

- Remote-sensing, Topography: newworldencyclopedia.org, Retrieved 30 March 2018

- Schott, John Robert (2007). Remote sensing: the image chain approach (2nd ed.). Oxford University Press. p. 1. ISBN 978-0-19-517817-3

- Twiss, S.D.; et al. (2001). "Topographic spatial characterisation of grey seal Halichoerus grypus breeding habitat at a sub-seal size spatial grain". Ecography. 24 (3): 257–266. doi:10.1111/j.1600-0587.2001.tb00198.x

- What-is-map: mapsofindia.com, Retrieved 20 April 2018

- The Remarkable History of GIS - Geographical Information Systems."The Remarkable History of GIS". Retrieved 2015-05-05

- Chang, K. T. (1989). "A comparison of techniques for calculating gradient and aspect from a gridded digital elevation model". International Journal of Geographical Information Science. 3 (4): 323–334. doi:10.1080/02693798908941519

- What-is-gps: circuitdigest.com, Retrieved 14 June 2018

4

The Atmosphere

The layer of gases surrounding a planet, which are held by the force of gravity is called the atmosphere. An understanding of the Earth's atmosphere is facilitated through the study of atmospheric composition, atmospheric layers, atmospheric pressure, climate change, atmospheric temperature, atmospheric circulation and the greenhouse effect, which have been thoroughly discussed in this chapter.

Atmospheric Composition

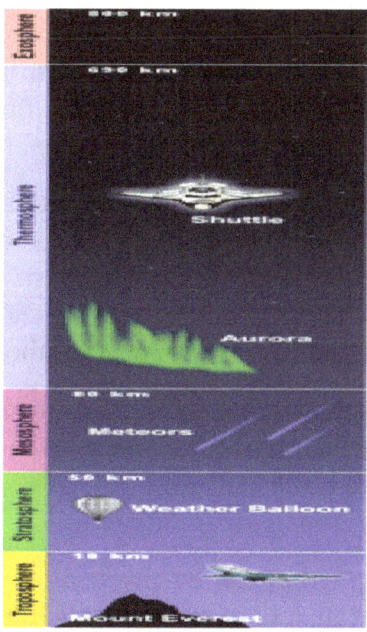
Layers of Atmosphere

The Earth's atmosphere is a layer of gases surrounding our planet and retained by the Earth's gravity. It contains roughly 78 percent nitrogen and 21 percent oxygen, with small amounts of carbon dioxide, water vapor, and other gases. This mixture of gases is commonly known as air. Based on its properties, the atmosphere is divided into several layers, but it has no abrupt, outermost boundary. It slowly becomes thinner and fades away into outer space.

The atmosphere protects and sustains life on Earth in a variety of ways. It provides oxygen for respiration, carbon dioxide for photosynthesis, nitrogen for nitrogen fixation, and water vapor for precipitation that nourishes the land. Carbon dioxide and water vapor reduce temperature extremes between day and night, keeping the planet warm enough for living organisms. The ozone layer absorbs ultraviolet solar radiation that could damage living tissue, and higher layers protect the Earth from bombardment by meteorites and charged particles in the solar wind.

One area of concern, though, is that human activities such as fuel burning and industrial production have been releasing pollutants into the atmosphere. In response, the governments of various nations have instituted measures to reduce the emission of pollutants.

Atmospheric Layers

The properties of the Earth's atmosphere vary with altitude. Based on these properties, the atmosphere may be regarded as having different layers or zones. According to one system of nomenclature, there are five layers: the troposphere, stratosphere, mesosphere, thermosphere, and exosphere. The boundaries between these regions are called the tropopause, stratopause, mesopause, and exobase.

Atmospheric regions are also named in other ways, as follows:

- Ozone layer (Ozonosphere): In the stratosphere, in an altitude range of about 10–50 km, the concentration of ozone (O_3) is a few parts per million, which is much higher than the ozone concentration in the lower atmosphere (although it is still small compared to the main components of the atmosphere). This layer, known as the ozone layer, is vitally important to life because it absorbs biologically harmful UV radiation from the Sun. Moreover, the absorbed solar energy raises the temperature of this part of the atmosphere, creating a thermal barrier that helps trap the atmosphere below, preventing it from bleeding out into space.

- Ionosphere: This is the region of the atmosphere that contains ions (that form a "plasma"), created by the interaction of solar radiation with gas particles. The ionosphere overlaps with the mesosphere and thermosphere, going up to an altitude of 550 km. Its value in practical terms is that it enables the propagation of radio wave signals, which bounce off the ions and can be transmitted to distant places on the Earth.

- Magnetosphere: It is the region where the Earth's magnetic field interacts with the solar wind. Its inner boundary is the ionosphere, but it extends for tens of thousands of kilometers, with a long tail away from the Sun.

- Van Allen radiation belts: These are regions where charged particles (forming a plasma) from the solar wind are trapped by the Earth's magnetic field. When the belts "overload," particles strike the upper atmosphere and fluoresce, producing the effects known as the polar auroras. Qualitatively, there are two belts: an inner belt, consisting mostly of protons, and an outer belt, consisting mostly of electrons.

- Homosphere (or Turbosphere) and Heterosphere: The region below the turbopause (that is, below an altitude of about 100 km) is known as the homosphere or turbosphere, where the chemical constituents are well mixed and the composition of the atmosphere remains fairly uniform. The region above the turbopause is called the heterosphere, where, in the absence of mixing, the chemical composition of the atmosphere varies.

Pressure, Density, and Mass

- Atmospheric pressure (or barometric pressure) is a direct result of the weight of the air. It is highest at the Earth's surface and decreases with altitude. This is because air at the surface is compressed by the weight of all the air above it. Air pressure varies with location and time, because the amount (and weight) of air above the Earth varies with location and time.

- Atmospheric pressure drops by approximately 50 percent at an altitude of about 5 km. (In other words, about 50 percent of the total atmospheric mass is within the lowest 5 km). The average atmospheric pressure at sea level is about 101.3 kilopascals (about 14.7 pounds per square inch).

- The density of air at sea level is about 1.2 kg/m3, and it decreases as altitude increases.

- The average mass of the atmosphere is about 5,000 trillion metric tons.

Thickness of the Atmosphere

- 57.8 percent of the atmosphere is below the summit of Mount Everest.

- 72 percent of the atmosphere is below the common cruising altitude of commercial airliners (about 10,000 m or 32,800 ft).

- 99.99999 percent of the atmosphere is below the highest flight altitude of the aircraft X-15, which reached 354,300 ft (108 km) on August 22, 1963. Therefore, most of the atmosphere (99.9999 percent) is below 100 km, although in the rarified region above this there are auroras and other atmospheric effects.

- The atmosphere exists at altitudes of 1,000 km and higher, but it is so thin as to be considered nonexistent.

Composition of the Atmosphere

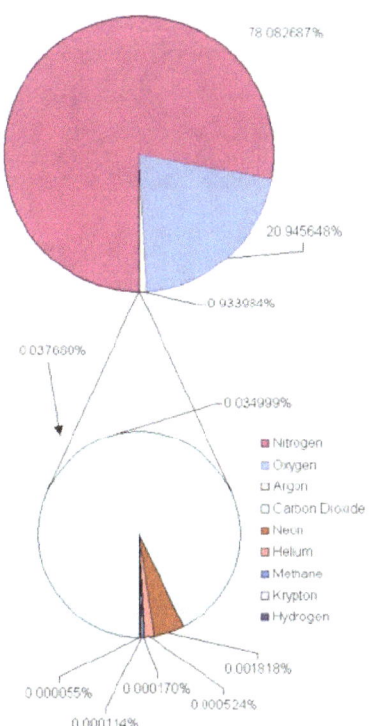

Composition of Earth's atmosphere. The lower pie represents the least common gases
that compose 0.038% of the atmosphere. Values normalized for illustration.

Composition of Dry Atmosphere (Homosphere), by Volume	
ppmv: Parts per million by volume	
Gas	**Volume**
Nitrogen (N_2)	780,840 ppmv (78.084%)
Oxygen (O_2)	209,460 ppmv (20.946%)
Argon (Ar)	9,340 ppmv (0.9340%)
Carbon dioxide (CO_2)	350 ppmv
Neon (Ne)	18.18 ppmv
Helium (He)	5.24 ppmv
Methane (CH_4)	1.745 ppmv
Krypton (Kr)	1.14 ppmv
Hydrogen (H_2)	0.55 ppmv
Not included in above dry atmosphere:	
Water vapor (highly variable)	Typically 1%

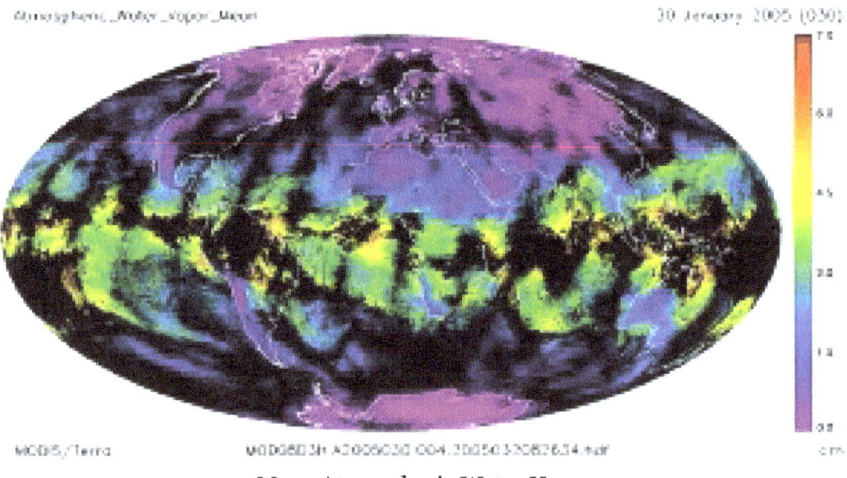

Mean Atmospheric Water Vapor.

Carbon dioxide and methane updated (to 1998) by IPCC TAR table. The NASA total was 17 ppmv over 100%, and CO_2 was increased here by 15 ppmv. To normalize, N_2 should be reduced by about 25 ppmv and O_2 by about 7 ppmv.

Minor components of air not listed above include:

Gas	Volume
Nitrous oxide	0.5 ppmv
Xenon	0.09 ppmv
Ozone	0.0 to 0.07 ppmv

Nitrogen dioxide	0.02 ppmv
Iodine	0.01 ppmv
Carbon monoxide	Trace
Ammonia	Trace

Biological Significance

The Earth's atmosphere plays a vital role in sustaining life on this planet. Oxygen is needed for respiration by animals, plants, and some bacteria. Nitrogen is an inert gas that reduces the amount of oxygen available for the oxidation of natural materials, thus restricting spontaneous combustion (burning) of flammable materials and the corrosion of metals. Nitrogen is also used by "nitrogen-fixing" bacteria to produce compounds that are useful for plant growth. Plants that perform photosynthesis take up carbon dioxide from the air and release oxygen. Carbon dioxide and water vapor act as "greenhouse gases" that keep the Earth sufficiently warm to maintain life. Water vapor in the air is part of the water cycle that produces precipitation (such as rain and snow) that replenishes moisture in the soil. In addition, water vapor prevents exposed living tissue from drying up.

Moreover, several regions of the atmosphere exert their protective effect from a distance. For instance, the ozone layer absorbs UV radiation that can damage the tissues and genetic material of living organisms. The mesosphere, in which millions of meteors burn up daily, protects the Earth's surface from being continually bombarded by these falling objects. The magnetosphere, which extends well beyond the atmosphere, protects the Earth from the damaging rain of charged particles carried by the solar wind.

Atmospheric Layers

The atmosphere is layered, corresponding with how the atmosphere's temperature changes with altitude.The atmosphere is divided into layers based on how the temperature in that layer changes with altitude, the layer's temperature gradient (Figure below). The temperature gradient of each layer is different. In some layers, temperature increases with altitude and in others it decreases. The temperature gradient in each layer is determined by the heat source of the layer (Figure below).

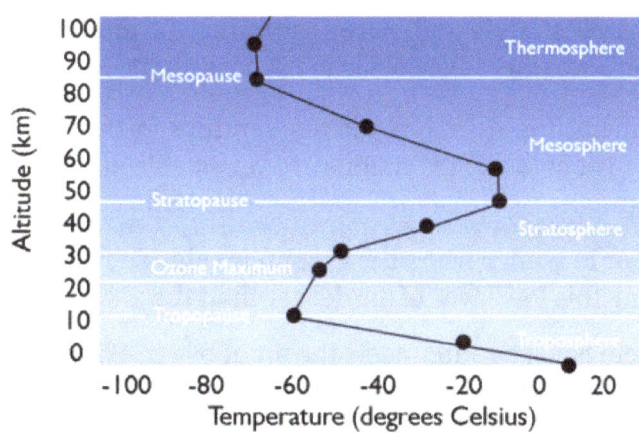

The four main layers of the atmosphere have different temperature gradients, creating the thermal structure of the atmosphere.

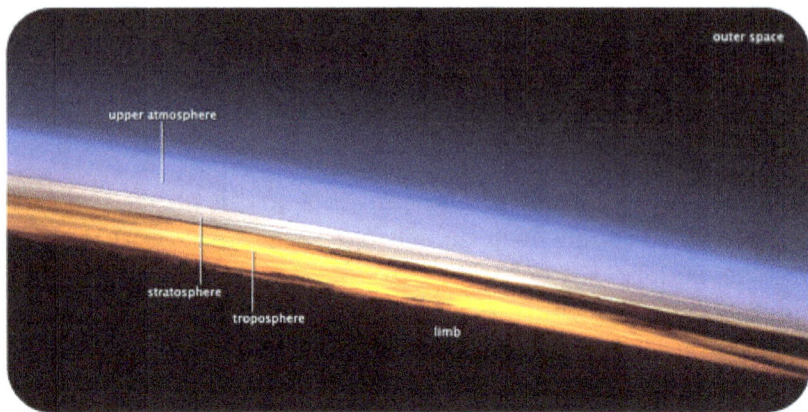

The layers of the atmosphere appear as different colors in this image from the International Space Station.

Most of the important processes of the atmosphere take place in the lowest two layers: the troposphere and the stratosphere.

Troposphere

The temperature of the troposphere is highest near the surface of the Earth and decreases with altitude. On average, the temperature gradient of the troposphere is 6.5°C per 1,000 m (3.6°F per 1,000 ft.) of altitude. What is the source of heat for the troposphere?

Earth's surface is a major source of heat for the troposphere, although nearly all of that heat comes from the Sun. Rock, soil, and water on Earth absorb the Sun's light and radiate it back into the atmosphere as heat. The temperature is also higher near the surface because of the greater density of gases. The higher gravity causes the temperature to rise.

Notice that in the troposphere warmer air is beneath cooler air. What do you think the consequence of this is? This condition is unstable. The warm air near the surface rises and cool air higher in the troposphere sinks. So air in the troposphere does a lot of mixing. This mixing causes the temperature gradient to vary with time and place. The rising and sinking of air in the troposphere means that all of the planet's weather takes place in the troposphere.

Sometimes there is a temperature inversion, air temperature in the troposphere increases with altitude and warm air sits over cold air. Inversions are very stable and may last for several days or even weeks. Inversions form:

- Over land at night or in winter when the ground is cold. The cold ground cools the air that sits above it, making this low layer of air denser than the air above it.

- Near the coast where cold seawater cools the air above it. When that denser air moves inland, it slides beneath the warmer air over the land.

Since temperature inversions are stable, they often trap pollutants and produce unhealthy air conditions in cities (Figure below).

Smoke makes a temperature inversion visible. The smoke is trapped in cold dense air that lies beneath a cap of warmer air.

At the top of the troposphere is a thin layer in which the temperature does not change with height. This means that the cooler, denser air of the troposphere is trapped beneath the warmer, less dense air of the stratosphere. Air from the troposphere and stratosphere rarely mix.

Stratosphere

Ash and gas from a large volcanic eruption may burst into the stratosphere, the layer above the troposphere. Once in the stratosphere, it remains suspended there for many years because there is so little mixing between the two layers. Pilots like to fly in the lower portions of the stratosphere because there is little air turbulence.

In the stratosphere, temperature increases with altitude. What is the heat source for the stratosphere? The direct heat source for the stratosphere is the Sun. Air in the stratosphere is stable because warmer, less dense air sits over cooler, denser air. As a result, there is little mixing of air within the layer.

The ozone layer is found within the stratosphere between 15 to 30 km (9 to 19 miles) altitude. The thickness of the ozone layer varies by the season and also by latitude.

The ozone layer is extremely important because ozone gas in the stratosphere absorbs most of the Sun's harmful ultraviolet (UV) radiation. Because of this, the ozone layer protects life on Earth. High-energy UV light penetrates cells and damages DNA, leading to cell death (which we know as a bad sunburn). Organisms on Earth are not adapted to heavy UV exposure, which kills or damages them. Without the ozone layer to reflect UVC and UVB radiation, most complex life on Earth would not survive long (Figure below).

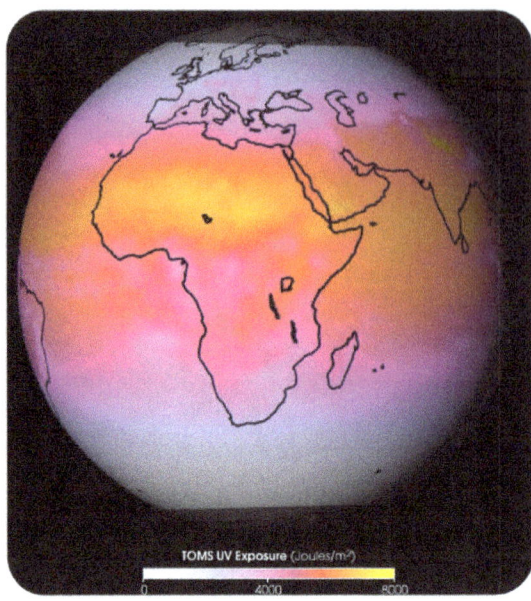

Even with the ozone layer, UVB radiation still manages to reach Earth's surface, especially where solar radiation is high.

Mesosphere

Temperatures in the mesosphere decrease with altitude. Because there are few gas molecules in the mesosphere to absorb the Sun's radiation, the heat source is the stratosphere below. The mesosphere is extremely cold, especially at its top, about -90°C (-130°F).

The air in the mesosphere has extremely low density: 99.9% of the mass of the atmosphere is below the mesosphere. As a result, air pressure is very low (Figure below). A person traveling through the mesosphere would experience severe burns from ultraviolet light since the ozone layer which provides UV protection is in the stratosphere below. There would be almost no oxygen for breathing. Stranger yet, an unprotected traveler's blood would boil at normal body temperature because the pressure is so low.

Meteors burn in the mesosphere even though the gas is very thin; these burning meteors are shooting stars.

The Thermosphere and Ionosphere

The thermosphere lies above the mesopause, and is a region in which temperatures again increase with height. This temperature increase is caused by the absorption of energetic ultraviolet and X-Ray radiation from the sun.

The region of the atmosphere above about 80 km is also caused the "ionosphere", since the energetic solar radiation knocks electrons off molecules and atoms, turning them into "ions" with a positive charge. The temperature of the thermosphere varies between night and day and between the seasons, as do the numbers of ions and electrons which are present. The ionosphere reflects and absorbs radio waves, allowing us to receive shortwave radio broadcasts in New Zealand from other parts of the world.

The Exosphere

The region above about 500 km is called the exosphere. It contains mainly oxygen and hydrogen atoms, but there are so few of them that they rarely collide - they follow "ballistic" trajectories under the influence of gravity, and some of them escape right out into space.

The Magnetosphere

The earth behaves like a huge magnet. It traps electrons (negative charge) and protons (positive), concentrating them in two bands about 3,000 and 16,000 km above the globe - the Van Allen "radiation" belts. This outer region surrounding the earth, where charged particles spiral along the magnetic field lines, is called the magnetosphere.

Atmospheric Pressure

The air around you has weight, and it presses against everything it touches. That pressure is called atmospheric pressure, or air pressure. It is the force exerted on a surface by the air above it as gravity pulls it to Earth.

Atmospheric pressure is commonly measured with a barometer. In a barometer, a column of mercury in a glass tube rises or falls as the weight of the atmosphere changes. Meteorologists describe the atmospheric pressure by how high the mercury rises.

An atmosphere (atm) is a unit of measurement equal to the average air pressure at sea level at a temperature of 15 degrees Celsius (59 degrees Fahrenheit). One atmosphere is 1,013 millibars, or 760 millimeters (29.92 inches) of mercury.

Atmospheric pressure drops as altitude increases. The atmospheric pressure on Denali, Alaska, is about half that of Honolulu, Hawai'i. Honolulu is a city at sea level. Denali, also known as Mount McKinley, is the highest peak in North America.

As the pressure decreases, the amount of oxygen available to breathe also decreases. At very high altitudes, atmospheric pressure and available oxygen get so low that people can become sick and even die.

Mountain climbers use bottled oxygen when they ascend very high peaks. They also take time to get used to the altitude because quickly moving from higher pressure to lower pressure can cause decompression sickness. Decompression sickness, also called "the bends", is also a problem for scuba divers who come to the surface too quickly.

Aircraft create artificial pressure in the cabin so passengers remain comfortable while flying.

Atmospheric pressure is an indicator of weather. When a low-pressure system moves into an area, it usually leads to cloudiness, wind, and precipitation. High-pressure systems usually lead to fair, calm weather.

A barometer measures atmospheric pressure, which is also called barometric pressure.

Atmospheric Temperature

The temperature of Earth's atmosphere varies with the distance from the equator (latitude) and height above the surface (altitude). It also changes with time, varying from season to season, and from day to night, as well as irregularly due to passing weather systems. If local variations are averaged out on a global basis, however, a pattern of global average temperatures emerges. Vertically, the atmosphere is divided into four layers: the troposphere, the stratosphere, the mesosphere, and the thermosphere.

The Vertical Temperature Profile

Averaging atmospheric temperatures over all latitudes and across an entire year gives us the average vertical temperature profile that is known as a standard atmosphere. The average vertical temperature profile suggests four distinct layers. In the first layer, known as the troposphere, average atmospheric temperature drops steadily from its value at the surface, about 290K (63°F; 17°C) and reaches of minimum of around 220K (−64°F;−53°C) at an altitude of about 6.2 mi (10 km). This level, known as the tropo-pause, is just above the cruising altitude of commercial jet aircraft. The decrease in temperature with height, called the lapse rate, is nearly steady throughout the troposphere at 43.7°F(6.5°C) per 0.6 mi (1 km). At the tropopause, the lapse rate abruptly decreases. Atmospheric temperature is nearly constant over the next 12 mi (20 km), then begins to rise with increasing altitude up to about 31 mi (50 km). This region of increasing temperatures is the stratosphere. At the top of the layer, called the stratopause, temperatures are nearly as warm as the surface values. Between about 31−50 mi (50−80 km) lies the mesosphere, where atmospheric temperature resumes its decrease with altitude and reaches a minimum of 180K (−136°F;−93°C) at the top of the layer (the mesopause), around 50 mi (80 km). Above the mesopause is the thermosphere that, as its name implies, is a zone of high gas temperatures. In the very high thermosphere (about 311 mi (500 km) above Earth's surface) gas temperatures can reach from 500−2,000K (441−3, 141°F; 227−1, 727°C). Temperature is a measure of the energy of the gas molecules' motion. Although they have high energy, the molecules in the thermosphere are present in very low numbers, less than one millionth of the amount present on average at Earth's surface.

Atmospheric temperature can also be plotted as a function of both latitude and altitude.

The Sun's Role in Atmospheric Temperature

Most solar radiation is emitted as visible light, with smaller portions at shorter wavelengths (ultraviolet radiation) and longer wavelengths (infrared radiation, or heat). Little of the visible light is absorbed by the atmosphere (although some is reflected back into space by clouds), so most of this energy is absorbed by Earth's surface. The Earth is warmed in the process and radiates heat (infrared radiation) back upward. This warms the atmosphere, and, just as one will be warmer when standing closer to a fire, the layers of air closest to the surface are the warmest.

According to this explanation, the temperature should continually decrease with altitude, however, shows that temperature increaseS with altitude in the stratosphere. The stratosphere contains nearly all the atmosphere's ozone. Ozone (O_3) and molecular oxygen (O_2) absorb most of the sun's short wavelength ultraviolet radiation. In the process they are broken apart and reform continuously. The net result is that the ozone molecules transform the ultraviolet radiation to heat energy, heating up the layer and causing the increasing temperature profile observed in the stratosphere.

The mesosphere resumes the temperature decrease with altitude. The thermosphere, however, is subject to very high energy, short wavelength ultraviolet and x-ray solar radiation. As the atoms or molecules present at this level absorb some of this energy, they are ionized (Have an electron removed) or dissociated (molecules are split into their component atoms). The gas layer is strongly heated by this energy bombardment, especially during periods when the sun is emitting elevated amounts of short wavelength radiation.

The Greenhouse Effect

Solar energy is not the only determinant of atmospheric temperature. As noted above, Earth's surface, after absorbing solar radiation in the visible region.

Emits infrared radiation back to space. Several atmospheric gases absorb this heat radiation and re-radiate it in all directions, including back toward the surface. These so-called greenhouse gases thus trap infrared radiation within the atmosphere, raising its temperature. Important greenhouse gases include water vapor (H_2O), carbon dioxide (CO_2), and methane (CH_4). It is estimated that Earth's surface temperature would average about 32°C (90°F) cooler in the absence of greenhouse gases. Because this temperature is well below the freezing point of water, the planet would be much less hos-pitable to life in the absence of the greenhouse effect.

While greenhouse gases are essential to life on the planet, more is not necessarily better. Since the beginning of the industrial revolution in the mid-nineteenth century, humans have released increasing amounts of carbon dioxide to the atmosphere by burning fossil fuels. The level of carbon dioxide measured in the remote atmosphere has shown a continuous increase since record keeping began in 1958. If this increase translates into a corresponding rise in atmospheric temperature, the results might include melting polar ice caps and swelling seas, resulting in coastal cities being covered by the ocean; shifts in climate perhaps leading to extinctions; and unpredictable changes in wind and weather patterns, posing significant challenges for agriculture. Predicting the changes that increased levels of greenhouse gases may bring is complicated. The interaction of the atmosphere, the oceans, the continents, and the ice caps is not completely understood. While it is known that some of the emitted carbon dioxide is absorbed by the oceans and eventually deposited as carbonate rock (such as limestone), it is not known if this is a steady process or if it can keep pace with current levels of carbon dioxide production.

Atmospheric Circulation

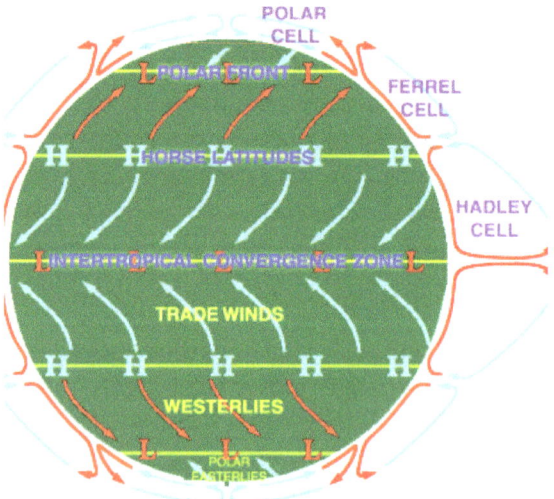

Atmospheric circulation cells and global windsystems

On Earth, an atmospheric circulation takes place (see picture) which is triggered by the temperature difference on the ground at the equator and poles. During the year, the sun is shining perpendicular at the equator whereby there is no sun in winter. In the summer, the sun only shines from a shallow angle. Thus, different pressure areas which trigger a large circulation between the equator and the poles are formed. Because of the earth´s rotation, a direct flow between anticyclone (equator) and depression (poles) is prevented. In the northern hemisphere, the air masses are defelcted to the right and in the southern hemisphere to the left. For that reason, three large circulation cells are generated (Hadley cell, Ferrel cell and the Polar cell).

The main effects of the atmospheric circulation:

- Continuous transport of humidity from the equator to the north and to the south tropics.

- Transport of hot air and humidity from the tropics to the temperate zones.

- Transport of warmer air and humidity from the temperate to the colder zones.

Hadley Cell

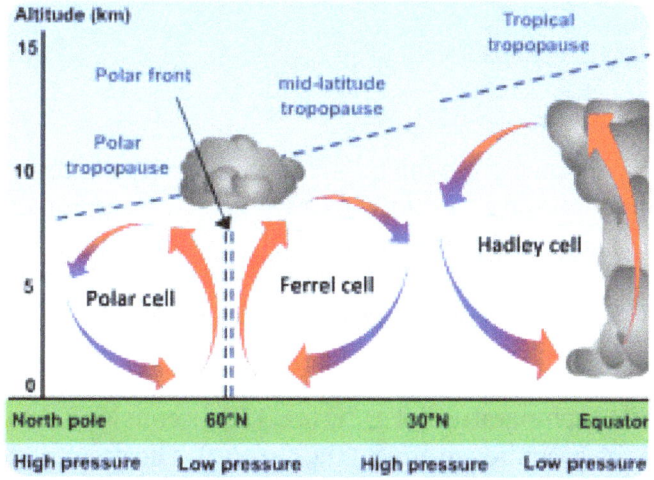

Atmospheric circulation cells

At the equator, the air rises up, because of strong heating by the sun. At the tropopause (temperature inversion in about 18km above ground), the air masses will deflect to the North and South. Through area correction, the air masses slide down to the poles. Furthermore, through the earth´s rotation, the winds fall until the 30th latitude and flow back to the equator as trade winds. At the equator, these winds meet in the Intertropical Convergence Zone (ITCZ). This circulation is called Hadley cell.

During the ascend process, the air cools down, the steam inside condenses, clouds are build and it starts to rain very strongly. In the descending process, the exact opposite is happening. The air gets warm and the water in it starts to evaporate. Desert areas (such as Sahara or Namib Desert) around the 30th latitude are consequences of this procedure. There are several anticyclones in this region, which is caused by the warm air on the ground. These are assigned to the subtropical ridge (horse latitudes).

Polar Cell

Near the ground level, air currents flow from the poles toward the equator. These are called polar easterlies, because they are distracted from eastside by the rotation of the earth. Near the 60th latitude, the winds are heated and rise up on the way to the South. This second circulation is called Polar cell.

Ferrel Cell

Based on air ascent (60th latitude) and air cooling (30th latitude), a third circulation is formed in the area between the 60th and 30th latitude. This circulation is called Ferrel cell. Near ground level, there is an air transport towards the poles wherey the air flows towards the equator at higher levels. In the northern hemisphere, the air on the ground is distracted to the right and in the southern hemisphere to the left. The winds from the West are called westerlies.

The polar front is located on the border between the polar easterlies (cold) and the easterlies (warm). This border is usually between the 60th and the 70th latitude. In this area, depressions often occure.

Greenhouse Effect

The greenhouse effect is the process in which long wave radiation (infrared) emitted by the earth surface is absorbed by atmospheric gases only to cause further emission of infrared radiation back to the earth, warming its surface. The major atmospheric gases causing such greenhouse effects are water vapor (H_2O), carbon dioxide (CO_2), methane (CH_4), ozone (O_3), nitrous oxide (N_2O), chlorofluorocarbons (CFCs), and perfluorocarbons (PFCs); they are known as greenhouse gases (GHGs).

The Earth's average surface temperature of 15°C (288 K) is considered to be about 33°C warmer than it would be without the greenhouse effect (IPCC 2007). The greenhouse effect was discovered by Joseph Fourier in 1824 and first investigated quantitatively by Swedish chemist Svante Arrhenius in 1896. Compared to the Earth, Mars shows very weak and Venus very strong greenhouse effects, as a result they have low and very high surface temperature, respectively.

The effect is derived from the greenhouse, as the warming of air inside a greenhouse compared to the air outside was supposed to take place in similar way.

The greenhouse effect is an important natural phenomenon allowing maintenance of a comfortable average temperature on the earth. A recent gradual warming of the Earth, generally known as global warming, is popularly considered to be the result of increased concentrations of greenhouse gases in the atmosphere as a result of human activities since the industrial revolution (Miller 2000), although there are divergent opinions among scientists regarding whether, or to what degree, temperature changes represent natural cycles or are anthropogenic in nature.

Basic Mechanism

The Earth receives energy from the Sun in the form a wide spectrum of electromagnetic radiation. However, over ninety percent of the Sun's radiation is in the form of visible light and infrared.

The Earth reflects about 30 percent of the incoming solar radiation; thus, the albedo (total reflectivity) value of the earth is 0.3. The remaining seventy percent is absorbed by atmosphere (19 percent), and by land and water (together 51 percent), warming the atmosphere, land, and oceans.

For the Earth's temperature to be in steady state so that the Earth does not rapidly heat or cool, the absorbed solar radiation must be very closely balanced by energy radiated back to space in the infrared wavelengths. Since the intensity of infrared radiation increases with increasing temperature, one can think of the Earth's temperature as being determined by the infrared radiation needed to balance the absorbed solar flux.

The visible solar radiation mostly heats the surface, not the atmosphere, whereas most of the infrared radiation escaping to space is emitted from the upper atmosphere, not the surface. Thirty percent of the solar flux is absorbed by the earth's surface and transferred to the atmosphere in the form of latent heat of vaporization during evaporation to be dissipated into the space as infrared waves. The remaining twenty one percent solar flux absorbed by the surface is emitted in the form of infrared photons; but they are mostly absorbed in the atmosphere by greenhouse gases and clouds and do not escape directly to space. The downward long–wave radiation occurs mostly from the atmosphere. This delayed dissipation of the solar flux due to greenhouse effect is responsible for warming effect.

The reason this warms the surface is most easily understood by starting with a simplified model of a purely radiative greenhouse effect that ignores energy transfer in the atmosphere by convection (sensible heat transport) and by the evaporation and condensation of water vapor (latent heat transport). In this purely radiative case, one can think of the atmosphere as emitting infrared radiation both upwards and downwards. The upward infrared flux emitted by the surface must balance not only the absorbed solar flux but also this downward infrared flux emitted by the atmosphere. The surface temperature will rise until it generates thermal radiation equivalent to the sum of the incoming solar and infrared radiation.

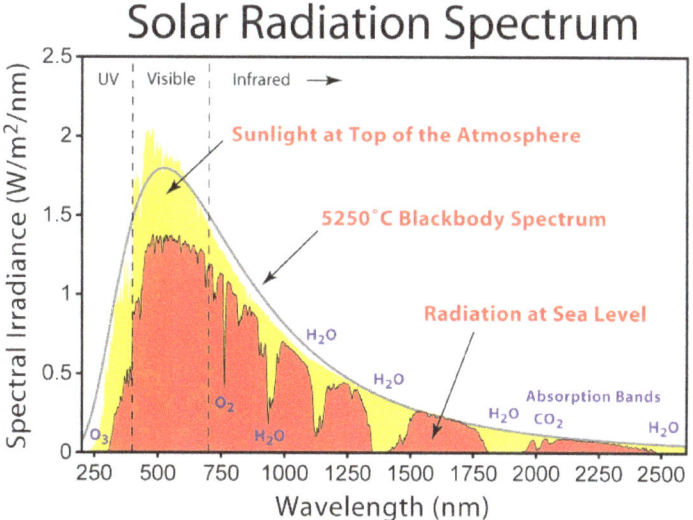

Solar radiation at top of atmosphere and at Earth's

A more realistic picture taking into account the convective and latent heat fluxes is somewhat more complex. But the following simple model captures the essence. The starting point is to note that the opacity of the atmosphere to infrared radiation determines the height in the atmosphere from which most of the photons are emitted into space. If the atmosphere is more opaque, the typical photon escaping to space will be emitted from higher in the atmosphere, because one then has to go to higher altitudes to see out to space in the infrared. Since the emission of infrared radiation is a function of temperature, it is the temperature of the atmosphere at this emission level that is effectively determined by the requirement that the emitted flux balance the absorbed solar flux.

But the temperature of the atmosphere generally decreases with height above the surface, at a rate of roughly 6.5°C per kilometer (km) on average, until one reaches the stratosphere 10−15 km above the surface. (Most infrared photons escaping to space are emitted by the troposphere, the region bounded by the surface and the stratosphere, so we can ignore the stratosphere in this simple picture.) A very simple model, but one that proves to be remarkably useful, involves the assumption that this temperature profile is simply fixed by the non−radiative energy fluxes. Given the temperature at the emission level of the infrared flux escaping to space, one then computes the surface temperature by increasing temperature at the rate of 6.5°C per kilometer, the environmental lapse rate, until one reaches the surface. The more opaque the atmosphere, and the higher the emission level of the escaping infrared radiation, the warmer the surface, since one then needs to follow this lapse rate over a larger distance in the vertical. While less intuitive than the purely radiative greenhouse effect, this less familiar radiative−convective picture is the starting point for most discussions of the greenhouse effect in the climate modeling literature.

The term "greenhouse effect" originally came from the greenhouses used for gardening. A greenhouse is built of glass, which is transparent to electromagnetic radiation in the visible part of the spectrum and not transparent to either side of the visible range (ultra violet and infrared). However, in reality the greenhouse heats up primarily because the Sun warms the ground inside it, which warms the air near the ground, and this air is prevented from rising and flowing away (Fraser). The warming inside a greenhouse thus occurs by suppressing convection and turbulent mixing. Greenhouses thus work primarily by preventing convection, just like the solar water heater. However, the atmospheric greenhouse effect of the Earth reduces radiation loss, not convection.

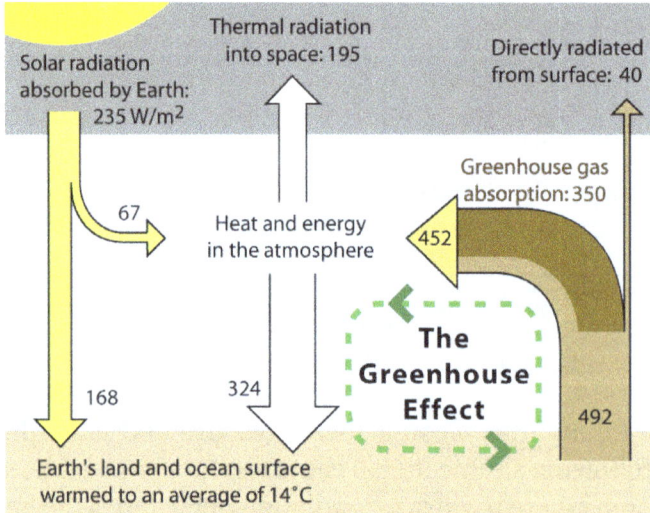

A schematic representation of the exchanges of energy between outer space, the Earth's atmosphere, and the Earth surface. The ability of the atmosphere to capture and recycle energy emitted by the Earth surface is the defining characteristic of the greenhouse effect.

A modern greenhouse in RHS Wisley

Greenhouse Gases

The molecules/atoms that constitute the bulk of the atmosphere—oxygen (O_2), nitrogen (N_2), and argon—do not interact with infrared radiation significantly. While the oxygen and nitrogen molecules can vibrate, because of their symmetry these vibrations do not create any transient charge separation. Without such a transient dipole moment, they can neither absorb nor emit infrared radiation.

In the Earth's atmosphere, the dominant infrared absorbing gases are water vapor, carbon dioxide, and ozone (O_3). The same molecules are also the dominant infrared emitting molecules.

Quantum mechanics provides the basis for computing the interactions between molecules and radiation. Most of this interaction occurs when the frequency of the radiation closely matches that of the spectral lines of the molecule, determined by the quantization of the modes of vibration and rotation of the molecule. Generally, the electronic excitations do not occur due to infrared radiation, as they require energy larger than that in an infrared photon. The width of a spectral line is an important element in understanding its importance for the absorption of radiation. In the Earth's atmosphere these spectral widths are primarily determined by "pressure broadening," which is the distortion of the spectrum due to the collision with another molecule. Most of the infrared absorption in the atmosphere can be thought of as occurring while two molecules are colliding. The absorption due to a photon interacting with a lone molecule is relatively small. This three–body aspect of the problem, one photon and two molecules, makes direct quantum mechanical computation for molecules of interest more challenging. Careful laboratory spectroscopic measurements, rather than ab initio quantum mechanical computations, provide the basis for most of the radiative transfer calculations used in studies of the atmosphere.

CO_2 and O_3 have "floppy" vibration motions whose quantum states can be excited by collisions at energies encountered in the atmosphere. For example, carbon dioxide is a linear molecule, but it has an important vibrational mode in which the molecule bends with the carbon in the middle moving one way and the oxygen atoms on the ends moving the other way, creating some charge separation, a dipole moment, and thus carbon dioxide molecules can absorb IR radiation.

Collisions will immediately transfer this energy to heating the surrounding gas. On the other hand, other CO2 molecules will be vibrationally excited by collisions. Roughly 5% of CO2 molecules are vibrationally excited at room temperature and it is this 5% that radiates. A substantial part of the greenhouse effect due to carbon dioxide exists because this vibration is easily excited by infrared radiation. CO2 has two other vibrational modes. The symmetric stretch does not radiate, and the asymmetric stretch is at too high a frequency to be effectively excited by atmospheric temperature collisions, although it does contribute to absorption of IR radiation.

The vibrational modes of water are at too high energies to effectively radiate, but do absorb higher frequency IR radiation. Water vapor has a bent shape. It has a permanent dipole moment (the O atom end is electron rich, and the H atoms electron poor) which means that IR light can be emitted and absorbed during rotational transitions (JEB 2002), and these transitions can also be produced by collisional energy transfer. Clouds are also very important infrared absorbers. Therefore, water has multiple effects on infrared radiation, through its vapor phase and through its condensed phases. Other absorbers of significance include methane, nitrous oxide and the chlorofluorocarbons.

Discussion of the relative importance of different infrared absorbers is confused by the overlap between the spectral lines due to different gases, widened by pressure broadening. As a result, the absorption due to one gas cannot be thought of as independent of the presence of other gases. One convenient approach is to remove the chosen constituent, leaving all other absorbers, and the temperatures, untouched, and monitoring the infrared radiation escaping to space. The reduction in infrared absorption is then a measure of the importance of that constituent. More precisely, one could define the greenhouse effect (GE) to be the difference between the infrared radiation that the surface would radiate to space if there were no atmosphere and the actual infrared radiation escaping to space. Then, one could compute the percentage reduction in GE when a constituent is removed. The table below is computed by this method, using a particular one–dimensional model (GISS–GCM ModelE) of the atmosphere (Lindzen 1991). More recent 3D computations lead to similar results.

Gas removed	percent reduction in GE
H_2O	36 percent
CO_2	9 percent
O_3	3 percent

By this particular measure, water vapor can be thought of as providing 36 percent of the greenhouse effect, and carbon dioxide 9 percent, but the effect of removal of both of these constituents will be greater than 48 percent. An additional proviso is that these numbers are computed holding the cloud distribution fixed. But removing water vapor from the atmosphere while holding clouds fixed is not likely to be physically relevant. In addition, the effects of a given gas are typically non–linear in the amount of that gas, since the absorption by the gas at one level in the atmosphere can remove photons that would otherwise interact with the gas at another altitude. The kinds of estimates presented in the table, while often encountered in the controversies surrounding global warming, must be treated with caution. Different estimates found in different sources typically result from different definitions and do not reflect uncertainties in the underlying radiative transfer.

Positive Feedback and Runaway Greenhouse Effect

When there is a loop of effects, such as the concentration of a greenhouse gas itself being a function of temperature, there is a feedback. If the effect is to act in the same direction on temperature, it is a positive feedback, and if in the opposite direction it is a negative feedback. Sometimes feedback effects can be on the same cause as the forcing but it can also be via another greenhouse gas or on other effects, such as change in ice cover affecting the planet's albedo.

Positive feedbacks do not have to lead to a runaway effect. With radiation from the Earth increasing in proportion to the fourth power of temperature, the feedback effect has to be very strong to cause a runaway effect. An increase in temperature from greenhouse gases leading to increased water vapor, which is a greenhouse gas, causing further warming is a positive feedback. This cannot be a runaway effect or the runaway effect would have occurred long ago. Positive feedback effects are common and can always exist while runaway effects are much rarer and cannot be operating at all times.

If the effects from the second iteration of the loop of effects is larger than the effects of the first iteration of the loop this will lead to a self perpetuating effect. If this occurs and the feedback only ends after producing a major temperature increase, it is called a runaway greenhouse effect. A runaway feedback could also occur in the opposite direction leading to an ice age. Runaway feedbacks are bound to stop, since infinite temperatures are not observed. They are allowed to stop due to things like a reducing supply of a greenhouse gas, or a phase change of the gas, or ice cover reducing towards zero or increasing toward a large size that is difficult to increase.

The runaway greenhouse effect could also be caused by liberation of methane gas from hydrates by global warming if there are sufficient hydrates close to unstable conditions. It has been speculated that the Permian–Triassic extinction event was caused by such a runaway effect. It is also thought that larger area of heat absorbing black soil could be exposed as the permafrost retreats and large quantities of methane could be released from the Siberian tundra as it begins to thaw, methane being 25 times more potent a greenhouse gas than carbon dioxide.

A runaway greenhouse effect involving CO_2 and water vapor may have occurred on Venus. On Venus today there is little water vapor in the atmosphere. If water vapor did contribute to the warmth of Venus at one time, this water is thought to have escaped to space. Venus is sufficiently strongly heated by the Sun that water vapor can rise much higher in the atmosphere and is split into hydrogen and oxygen by ultraviolet light. The hydrogen can then escape from the atmosphere and the oxygen recombines. Carbon dioxide, the dominant greenhouse gas in the current atmosphere of Venus, likely owes its larger concentration to the weakness of carbon recycling as compared to Earth, where the carbon dioxide emitted from volcanoes is efficiently subducted into the Earth by plate tectonics on geologic time scales.

Anthropogenic Greenhouse Effect

Because of the greenhouse effect, a significant increase in greenhouse gases should translate to increase in global mean temperature. Currently, there is a view among many scientists and layman that there is indeed an increase in globally averaged temperatures since the mid-20th century and that it is most likely a result of an observed increase in anthropogenic greenhouse gas concentrations (IPCC 2007). However, other scientists and layman contend that present temperature increases are part of a natural cycle of temperature fluctuations, seen throughout geologic history, and not part of anthropogenic effects, and that carbon dioxide levels have not increased enough to make a significant temperature difference. A seemingly smaller group of scientists contend that there is not even a consistent increase in global mean temperatures, but observed increases are an artifact of the way temperatures are measured.

The ice core data from over the past 800,000 years does show that carbon dioxide has varied from values as low as 180 parts per million (ppm) to the pre–industrial level of 270 ppm. Measurements of carbon dioxide amounts from Mauna Loa observatory show that CO_2 has increased from about 313 ppm (parts per million) in 1960 to about 380 ppm in 2005 (Hileman 2005). The current concentration of CO_2 is 27% higher than the pre–industrial level and is higher than that of any time in the last 800,000 years history of the earth (Amos 2006). CO_2 production from increased industrial activity (fossil fuel burning) and other human activities such as cement production, biomass burning, and tropical deforestation has increased the CO_2 concentrations in the atmosphere.

Certain paleoclimatologists consider variations in carbon dioxide to be a fundamental factor in controlling climate variations over this time scale (Browen 2005). However, other greenhouse gases like CFCs, methane, and nitrous oxide have also risen substantially in the recent decade (Miller 2000). Methane is produced when methanogenic bacteria utilize organic matter in moist places that lack oxygen. The most favorable sites of methane productions are swamps and other natural wetlands, paddy fields, landfills, as well as the intestines of ruminants, termites, and so forth. CFCs are already banned, but the previously introduced enormous amount is still active. Nitrous oxide is released in the atmosphere from burning biomass, nitrogen rich fossil fuel (especially the coal), nylon production, denitrification process in organic substance and nitrate containing anaerobic soils and water bodies. Although molecules of CFCs, methane, and nitrous oxide absorbs and radiate much more infrared per molecule than CO_2, the much larger input of CO_2 makes it the most important greenhouse gas produced by human activities.

However, it should be noted that temperatures have cycled significantly during geologic history and even in the past 800,000 years, such as the Younger Dryas (10,000–11,000 BP, a time of relatively abrupt cold climate conditions); Holocene Climatic Optimum (Holocene thermal maximum ~7000 BP–3000 BP); Medieval Warm Period (900–1300 C.E.); Little Ice Age (1300–1800 C.E.), and Year without a summer (1816 C.E.). Some scientists contend that there is presently a warming period but that it is part of such normal cycles and not a result of an increase in greenhouse gases.

Climate Change

The planet's climate has constantly been changing over geological time. The global average temperature today is about 15C, though geological evidence suggests it has been much higher and lower in the past.

However, the current period of warming is occurring more rapidly than many past events. Scientists are concerned that the natural fluctuation, or variability, is being overtaken by a rapid human-induced warming that has serious implications for the stability of the planet's climate.

Climate change (or global warming), is the process of our planet heating up.

The Earth has warmed by an average of 1°C in the last century, and although that might not sound like much, it means big things for people and wildlife around the globe.

Unfortunately, rising temperatures don't just mean that we'll get nicer weather – if only! The changing climate will actually make our weather more extreme and unpredictable.

As temperatures rise, some areas will get wetter and lots of animals (and humans!) could find they're not able to adapt to their changing climate.

Causes of Climate Change

1. Burning fossil fuels

Over the past 150 years, industrialised countries have been burning large amounts of fossil fuels such as oil and gas. The gases released into the atmosphere during this process act like an invisible 'blanket', trapping heat from the sun and warming the Earth. This is known as the "Greenhouse Effect".

2. Farming

Believe it or not, cows' eating habits contribute towards greenhouse gases. Just like us, when cows eat, methane gas builds up in their digestive system and is released in the form of a fart. This might sound funny, but when you imagine that there are almost 1.5 billion cows releasing all that gas into the atmosphere, it sure adds up!

3.Deforestation

Forests absorb huge amounts of carbon dioxide – a greenhouse gas – from the air, and release oxygen back into it. The Amazon rainforest is so large and efficient at doing this that it is often called 'the lungs of the Earth'. Sadly, many rainforests are being cut down to make wood, palm oil and to clear the way for farmland, roads, oil mines, and dams.

Affect of Climate Change on Planet

The Earth has had many tropical climates and ice ages over the billions of years that it's been in existence, so why is now so different? Well, this is because for the last 150 years human activity has meant we're releasing a huge amount of harmful gases into the Earth's atmosphere, and records show that the global temperatures are rising more rapidly since this time.

A warmer climate could affect our planet in a number of ways:

- More rainfall

- Changing seasons

- Shrinking sea ice

- Rising sea levels

Affect of Climate Change on Wildlife

Climate change is already affecting wildlife all over the world, but certain species are suffering more than others. Polar animals – whose icy natural habitat is melting in the warmer temperatures – are particularly at risk. In fact, experts believe that the Arctic sea ice is melting at a shocking rate – 9% per decade! Polar bears need sea ice to be able to hunt, raise their young and as places to rest after long periods of swimming. Certain seal species, like ringed seals make caves in the snow and ice to raise their pups, feed and mate.

It's not just polar animals who are in trouble. Apes like orangutans, which live in the rainforests of Indonesia, are under threat as their habitat is cut down, and more droughts cause more bushfires.

Sea turtles rely on nesting beaches to lay their eggs, many of which are threatened by rising sea levels. Did you know that the temperature of nests determines whether the eggs are male or female? Unfortunately, with temperatures on the rise, this could mean that many more females are born than males, threatening future turtle populations.

Affect of Climate Change on People

Climate change won't just affect animals, it's already having an impact on people, too. Most affected are some of the people who grow the food we eat every day. Farming communities, especially in developing countries, are facing higher temperatures, increased rain, floods and droughts.

We Brits love a good cuppa, (around165 million cups of the stuff every day!), but we probably take for granted just how much work goes into growing our tea. Environmental conditions can affect the flavour and quality plus it needs a very specific rainfall to grow. In Kenya, climate change is making rainfall patterns less and less predictable. Often there will be droughts followed by huge amounts rain, which makes it very difficult to grow tea.

Farmers might then resort to using cheap chemicals to improve their crop to earn more money, even when long-term use of these chemicals can destroy their soil.

References

- Atmosphere, Earth: newworldencyclopedia.org, Retrieved 22 June 2018

- Atmospheric-layers: lumenlearning.com, Retrieved 12 July 2018

- Atmospheric-pressure, encyclopedia: nationalgeographic.org, Retrieved 25 March 2018

- Atmospheric-temperature, encyclopedias-almanacs-transcripts-and-maps, science: encyclopedia.com, Retrieved 15 May 2018

- Greenhouse-effect: newworldencyclopedia.org, Retrieved 25 April 2018

- What-is-climate-change: general-geography: natgeokids.com, Retrieved 14 July 2018

The Biosphere

The biosphere is an ecological system comprising of living organisms and their interactions with each other and with the spheres of the Earth. The topics elaborated in this chapter will help in developing an insight into the biosphere, life on Earth, species biodiversity, ecological succession, ecological pyramid and the biogeochemical cycle.

Life on Earth

Life! It's everywhere on Earth; you can find living organisms from the poles to the equator, from the bottom of the sea to several miles in the air, from freezing waters to dry valleys to undersea thermal vents to groundwater thousands of feet below the Earth's surface. Over the last 3.7 billion years or so, living organisms on the Earth have diversified and adapted to almost every environment imaginable. The diversity of life is truly amazing, but all living organisms do share certain similarities. All living organisms can replicate, and the replicator molecule is DNA. As well, all living organisms contain some means of converting the information stored in DNA into products used to build cellular machinery from fats, proteins, and carbohydrates.

Three Domains of Life

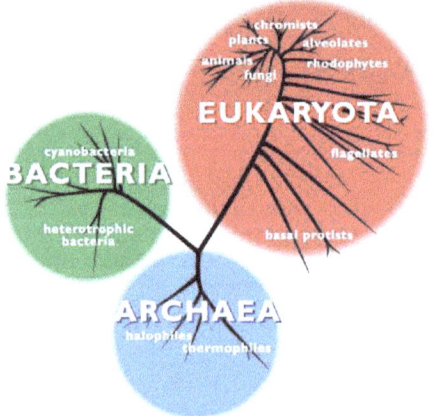

Until comparatively recently, living organisms were divided into two kingdoms: animal and vegetable, or the Animalia and the Plantae. In the 19th century, evidence began to accumulate that these were insufficient to express the diversity of life, and various schemes were proposed with three, four, or more kingdoms. The scheme most often used currently divides all living organisms into five kingdoms: Monera (bacteria), Protista, Fungi, Plantae, and Animalia. This coexisted with a scheme dividing life into two main divisions: the Prokaryotae (bacteria, etc.) and the Eukaryotae (animals, plants, fungi, and protists).

Recent work, however, has shown that what were once called "prokaryotes" are far more diverse than anyone had suspected. The Prokaryotae are now divided into two domains, the Bacteria and the Archaea, as different from each other as either is from the Eukaryota, or eukaryotes. No one of these groups is ancestral to the others, and each shares certain features with the others as well as having unique characteristics of its own.

Within the last two decades, a great deal of additional work has been done to resolve relationships within the Eukaryota. It now appears that most of the biological diversity of eukaryotes lies among the protists, and many scientists feel it is just as inappropriate to lump all protists into a single kingdom as it was to group all prokaryotes. Although many revised systems have been proposed, no single one of them has yet gained a wide acceptance.

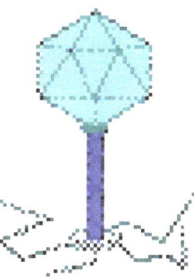

A fourth group of biological entities, the viruses, are not organisms in the same sense that eukaryotes, archaeans, and bacteria are. However, they are of considerable biological importance.

In all cladograms in our exhibits, if there is a picture within a box, that means we have an exhibit on the taxon. If your favorite organisms aren't here yet, keep trying: since there may be as many as 100 million living and fossil species of organism, it may take us a little while to cover all the highlights.

Abiogenesis

Abiogenesis is the idea that life arose from nonlife more than 3.5 billion years ago on Earth.

Abiogenesis proposes that the first life-forms generated were very simple and through a gradual process became increasingly complex. Biogenesis, in which life is derived from the reproduction of other life, was presumably preceded by abiogenesis, which became impossible once Earth's atmosphere assumed its present composition.

Although many equate abiogenesis with the archaic theory of spontaneous generation, the two ideas are quite different. According to the latter, complex life (e.g., a maggot or mouse) was thought to arise spontaneously and continually from nonliving matter. While the hypothetical process of spontaneous generation was disproved as early as the 17th century and decisively rejected in the 19th century, abiogenesis has been neither proved nor disproved.

The Oparin-Haldane Theory

In the 1920s British scientist J.B.S. Haldane and Russian biochemist Aleksandr Oparin independently set forth similar ideas concerning the conditions required for the origin of life on Earth.

Both believed that organic molecules could be formed from abiogenic materials in the presence of an external energy source (e.g., ultraviolet radiation) and that the primitive atmosphere was reducing (having very low amounts of free oxygen) and contained ammonia and water vapour, among other gases. Both also suspected that the first life-forms appeared in the warm, primitive ocean and were heterotrophic (obtaining preformed nutrients from the compounds in existence on early Earth) rather than autotrophic (generating food and nutrients from sunlight or inorganic materials).

Aleksandr Oparin, 1970.*Tass/Sovfoto*

Oparin believed that life developed from coacervates, microscopic spontaneously formed spherical aggregates of lipid molecules that are held together by electrostatic forces and that may have been precursors of cells. Oparin's work with coacervates confirmed that enzymes fundamental for the biochemical reactions of metabolism functioned more efficiently when contained within membrane-bound spheres than when free in aqueous solutions. Haldane, unfamiliar with Oparin's coacervates, believed that simple organic molecules formed first and in the presence of ultraviolet light became increasingly complex, ultimately forming cells. Haldane and Oparin's ideas formed the foundation for much of the research on abiogenesis that took place in later decades.

The Miller-Urey Experiment

In 1953 American chemists Harold C. Urey and Stanley Miller tested the Oparin-Haldane theory and successfully produced organic molecules from some of the inorganic components thought to have been present on prebiotic Earth. In what became known as the Miller-Urey experiment, the two scientists combined warm water with a mixture of four gases—water vapour, methane, ammonia, and molecular hydrogen—and pulsed the "atmosphere" with electrical discharges. The different components were meant to simulate the primitive ocean, the prebiotic atmosphere, and heat (in the form of lightning), respectively. One week later Miller and Urey found that simple organic molecules, including amino acids (the building blocks of proteins), had formed under the simulated conditions of early Earth.

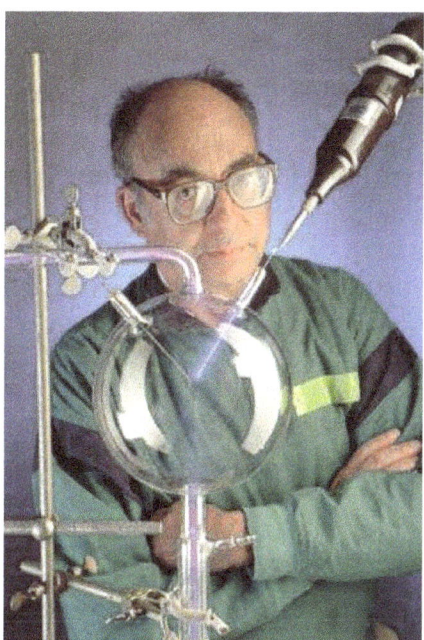

Miller, Stanley; Miller-Urey experimentAmerican chemist Stanley Miller with the components of the 1953 experiment he and Harold C. Urey undertook.Jim Sugar/Corbis

Modern Conceptions of Abiogenesis

Modern abiogenesis hypotheses are based largely on the same principles as the Oparin-Haldane theory and the Miller-Urey experiment. There are, however, subtle differences between the several models that have been set forth to explain the progression from abiogenic molecule to living organism, and explanations differ as to whether complex organic molecules first became self-replicating entities lacking metabolic functions or first became metabolizing protocells that then developed the ability to self-replicate.

The habitat for abiogenesis has also been debated. While some evidence suggests that life may have originated from nonlife in hydrothermal vents on the ocean floor, it is possible that abiogenesis occurred elsewhere, such as deep below Earth's surface, where newly arisen protocells could have subsisted on methane or hydrogen, or even on ocean shores, where proteinoids may have emerged from the reaction of amino acids with heat and then entered the water as cell-like protein droplets.

Some scientists have proposed that abiogenesis occurred more than once. In one example of this hypothetical scenario, different types of life arose, each with distinct biochemical architectures reflecting the nature of the abiogenic materials from which they developed. Ultimately, however, phosphate-based life ("standard" life, having a biochemical architecture requiring phosphorus) gained an evolutionary advantage over all non-phosphate-based life ("nonstandard" life) and thereby became the most widely distributed type of life on Earth. This notion led scientists to infer the existence of a shadow biosphere, a life-supporting system consisting of microorganisms of unique or unusual biochemical structure that may have once existed, or possibly still exists, on Earth.

As the Miller-Urey experiment demonstrated, organic molecules can form from abiogenic materials under the constraints of Earth's prebiotic atmosphere. Since the 1950s, researchers have found that amino acids can spontaneously form peptides (small proteins) and that key intermediates in the synthesis of RNA nucleotides (nitrogen-containing compounds [bases] linked to sugar and phosphate groups) can form from prebiotic starting materials. The latter evidence may support the RNA world hypothesis, the idea that on early Earth there existed an abundance of RNA life produced through prebiotic chemical reactions. In fact, in addition to carrying and translating genetic information, RNA is a catalyst, a molecule that increases the rate of a reaction without itself being consumed, meaning that a single RNA catalyst could have produced multiple living forms, which would have been advantageous during the rise of life on Earth. The RNA world hypothesis is one of the leading self-replication-first conceptions of abiogenesis.

Some modern metabolism-based models of abiogenesis incorporate Oparin's enzyme-containing coacervates but suggest a steady progression from simple organic molecules to coacervates, specifically protobionts, aggregates of organic molecules that display some characteristics of life. Protobionts presumably then gave rise to prokaryotes, single-celled organisms lacking a distinct nucleus and other organelles because of the absence of internal membranes but capable of metabolism and self-replication and susceptible to natural selection. Examples of primitive prokaryotes still found on Earth today include archaea, which often inhabit extreme environments with conditions similar to those that may have existed billions of years ago, and cyanobacteria (blue-green algae), which also flourish in inhospitable environments and are of particular interest in understanding the origin of life, given their photosynthetic abilities. Stromatolites, deposits formed by the growth of blue-green algae, are the world's oldest fossils, dating to 3.5 billion years ago.

Blue-green algae in Morning Glory Pool, Yellowstone National Park, Wyoming.

There remain many unanswered questions concerning abiogenesis. Experiments have yet to demonstrate the complete transition of inorganic materials to structures like protobionts and protocells and, in the case of the proposed RNA world, have yet to reconcile important differences in mechanisms in the synthesis of purine and pyrimidine bases necessary to form complete RNA nucleotides. In addition, some scientists contend that abiogenesis was unnecessary, suggesting instead that life was introduced on Earth via collision with an extraterrestrial object harbouring living organisms, such as a meteorite carrying single-celled organisms; the hypothetical migration of life to Earth is known as panspermia.

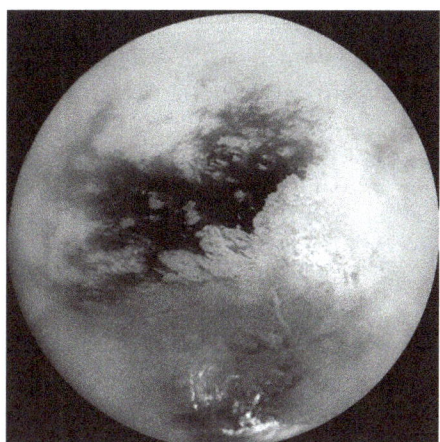

Saturn's moon Titan, in a mosaic of nine images taken by the Cassini spacecraft,
and processed to reduce the veiling effects of the moon's atmosphere.

The view is centred slightly south of the equator, with north toward the top. The continent-size region Xanadu Regio shows as the large bright patch on the right, while bright methane clouds appear near Titan's south pole.

Research on abiogenesis has benefited significantly from astrobiology, the field of study concerned with the search for extraterrestrial life (life beyond Earth) and with understanding the conditions required for life to form. Astrobiological investigations of the moon Titan, for example, which has an atmosphere lacking free oxygen, have revealed that complex organic molecules are present there, offering scientists a glimpse into the formation of biological materials in a prebiotic habitat resembling that of early Earth.

Biogenesis

Biogenesis refers to the production of life from life. The production of life from non-living matter is abiogenesis. Abiogenesis has never happened naturally and there are no accepted models for how it could happen in a laboratory or elsewhere. Biogenesis occurs constantly, from the division of bacteria, to trees shedding seeds, to cows calving.

Biogenesis is a relatively new understanding of how life reproduces, as people as late as 150 years ago believed that life came from inanimate materials. For example, because mice and flies were abundant near straw and meat, people thought they came from straw or meat. Louis Pasteur proved spontaneous generation was false in the 1860s. Now, scientists see biogenesis is the only natural source of living things.

Species Biodiversity

Species diversity is a term used to describe the measurement of biological diversity to be found in a specific ecological community. It includes the species richness or number of species to be found in an ecological community, the abundance (or number of individuals per species) and distribution or evenness of species.

Species diversity as compared to benchmarks can be used to evaluate the health of ecosystems. In a healthy ecosystem, the species that would ordinarily occupy various niches in an untouched, benchmark site would be present in similar richness, abundance and evenness. Species diversity measurements can be used to determine the need for conservation interventions and a high level of species richness may indicate the presence of rare species or species that are unique to an ecosystem.

Significance of Species Diversity

The composition of species in a given ecosystem is the result of longlasting evolution. Each species has adapted to its own niche, which is characterized by certain features (e.g. temperature range, availability of food or light) enabling the species to reproduce and thus maintain its population.

Living in an ecosystem, the species interacts with its environment (e.g. mussels take particles out of the water, reed forms root systems) and thus performs certain functions (increasing the light availability for plant growth, preventing sediment erosion). In a natural state, these interactions and consequently the system is in balance.

The loss of one species affects many other species and causes imbalance. As a result, several functions within and of the system are not carried out any more. Any species that will take over the lost specie's niche will most certainly not replace all of the functions it used to perform.

When species get extinct, their services for the global biosphere are lost for ever. It is impossible to replace it.

Affect of Human Activities on Species Diversity

Over-exploitation, pollution and habitat conversion are the main threats to species diversity. They cause a gradual loss of species on local, regional and global levels. Additionally, the introduction of species into new ecosystems destroys natural balance.

The ever-growing tendencies of tourism, transport, profit-oriented food production (e.g. single-crop agriculture, selective (?) aquaculture), and industry enforce these human activities.

Global warming and population growth continually increase these pressures on biodiversity.

Ecological Succession

Ecological succession is a fundamental concept in ecology, refers to more-or-less predictable and orderly changes in the composition or structure of an ecological community, resulting from biotic changes in resource supply. This process is governed by competition for resources. Succession may be initiated either by formation of new, unoccupied habitat (*e.g.*, a lava flow or a severe landslide) or by some form of disturbance (*e.g.* fire, severe windthrow, logging) of an existing community. The former case is often referred to as *primary succession*, the latter as *secondary succession*.

Eugene Odum compared succession to the development or maturation of an organism, and regarded the view that "ecological succession is a developmental process and not just a succession of species each acting alone" as "one of the most important unifying theories in ecology" (Odum 1983; Goldsmith 1985). Such a concept highlights the fact that ecosystems, just like individual organisms, develop through an orderly procession of stages, with subsequent stages dependent on those preceding.

The trajectory of ecological change can be influenced by site conditions, by the interactions of the species present, and by more stochastic factors such as availability of colonists, or seeds, or weather conditions at the time of disturbance. Some of these factors contribute to predictability of successional dynamics; others add more probabilistic elements. In general, communities in early succession will be dominated by fast-growing, well-dispersed species (opportunist, fugitive, or *r-selected* life-histories). As succession proceeds, these species will tend to be replaced by more competitive (*k-selected*) species. Typically, r-selected species produce many offspring, each of which is unlikely to survive to adulthood, while K-selected species invest more heavily in fewer offspring, each of which has a better chance of surviving to adulthood.

Trends in ecosystem and community properties in succession have been suggested, but few appear to be general. For example, species diversity almost necessarily increases during early succession as new species arrive, but may decline in later succession as competition eliminates opportunistic species and leads to dominance by locally superior competitors. Net primary production, biomass, and trophic properties all show variable patterns over succession, depending on the particular system and site.

Secondary succession: trees are colonizing uncultivated fields and meadows

Primary Succession

Primary succession is the orderly and predictable series of events through which a stable ecosystem forms in a previously uninhabited region. Primary succession occurs in regions characterized by the absence of soil and living organisms.

It begins with the appearance of pioneer species – lichen, mosses and fungi – that can grow on rocks and exposed land. These are small, simple organisms that can survive harsh conditions, fix inorganic carbon and nitrogen, and accelerate the process of weathering. As they die and decompose, their organic matter becomes the foundation for a thin layer of soil. Pioneer species pave the way for more complex communities of organisms, because the pioneers have altered the physical environment to make it more habitable. Once grasses and weeds begin to grow, soil formation is accelerated and more animal species begin to appear. The environment retains moisture, and ideal conditions are created for the growth of shrubs and small trees. This is followed by larger trees and animals, and the complex web of interactions between them.

Examples of Primary Succession

Primary succession can occur after a variety of events. These include:

- Volcanic eruptions

- Retreat of glaciers

- Flooding accompanied by severe soil erosion

- Landslides

- Nuclear explosions

- Oil spills

- Abandonment of a manmade structure, such as a paved parking lot

While some of these are natural events, some are anthropogenic, or manmade.

Primary Succession after a Volcanic Eruption

Lava from an erupting volcano incinerates everything in its path and forms new land that is made from inorganic material. While it is rich in minerals, the land cannot support a varied and complex ecosystem. Its capacity to sustain a stable ecosystem is limited. Pioneer species that colonize areas after volcanic eruptions include swordfern and green algae.

Carrizozo Lava Flow

A few small invertebrate animals may also venture into this territory, followed by crickets and spiders.

In the case of volcanic eruptions in the ocean, the atolls formed are isolated from other terrestrial ecosystems and have unique food chains and webs. Pioneer species often arise from spores carried through ocean currents.

Primary Succession in Sand Dunes

Seashores are harsh environments because of high wind speeds, moving sand and the minimal availability of fresh water and organic nutrients. Pioneer plants in such environments tend to have symbiotic bacteria in their root nodules to fix nitrogen. They also have root systems that can anchor them in shifting sand, have multiple adaptations to harvest fresh water, and reduce water loss through transpiration. Examples of pioneer species in sand dunes include sand couch grass and lyme grass.

These are followed by other grasses, and then by lichens that are deposited on the thin layer of organic matter created by the pioneer species. As the ecosystem develops, bracken, gorse, heather, hawthorn and brambles can be seen.

Eventually, a woodland will develop, containing organisms that can thrive in a high salt environment.

Primary Succession after a Nuclear Explosion

Some islands in French Polynesia were used for extensive testing of nuclear bombs in the 1960s and 70s. They were completely denuded of all plant, animal and microbial life. Scientists estimated that it would take centuries before life returned to these islands. However, surveys conducted over the course of 30 years show that primary succession has begun, and many islands have grasses, mosses and some plants. Some species of mollusks have also begun to live on these islands.

After the major accident at Chernobyl Nuclear Reactor in Ukraine (1986), the area was evacuated and has had minimal human habitation for the past three decades. The central reactor is still highly radioactive and is considered a completely 'dead' zone. However, robots sent into the heart of this reactor returned with black fungi that were using the radiation itself as an energy source.

While the high radiation levels limit the scope of research into these ecosystems, it will be of great interest to continue studying primary succession in these environments.

Differences Between Primary and Secondary Succession

Secondary succession occurs after an event that deeply disturbs an existing, stable ecosystem when most above-ground vegetation and living organisms disappear from the region. Though it appears as if the region is 'dead', the soil remains fertile and contains enough organic matter to support the reappearance of life. Grasses are among the first species to appear, quickly followed by shrubs and small trees.

The major difference between primary and secondary succession is the quality of the soil. Secondary succession does not require pedogenesis or soil formation. For example, primary succession would occur on barren land that was previously covered by a glacier, while secondary succession would occur on land after a forest fire. The forest fire may destroy all the plants and drive away the animals, but the ashes and decomposing organic matter can enrich the soil, and life restarts from sprouting roots and shoots and through the germination of seeds already present in the soil. In the case of the retreating glacier, however, the land has not supported life for hundreds of thousands of years and lacks any organic matter.

Secondary Succession

Secondary succession is the series of community changes which take place on a previously colonized, but disturbed or damaged habitat. Examples include areas which have been cleared of existing vegetation (such as after tree-felling in a woodland) and destructive events such as fires.

Secondary succession is usually much quicker than primary succession for the following reasons:

- There is already an existing seed bank of suitable plants in the soil.

- Root systems undisturbed in the soil, stumps and other plant parts from previously existing plants can rapidly regenerate.

- The fertility and structure of the soil has also already been substantially modified by previous organisms to make it more suitable for growth and colonization.

Examples

Imperata

Imperata grasslands are caused by human activities such as logging, forest clearing for shifting cultivation, agriculture and grazing, and also by frequent fires. The latter is a frequent result of human interference. However, when not maintained by frequent fires and human disturbances, they regenerate naturally and speedily to secondary young forest. The time of succession in *Imperata* grassland (for example in Samboja Lestari area), *Imperata cylindrica* has the highest coverage but it becomes less dominant from the fourth year onwards. While *Imperata* decreases, the percentage of shrubs and young trees clearly increases with time. In the burned plots, *Melastoma malabathricum, Eupatorium inulaefolium, Ficus* sp., and *Vitex pinnata*. strongly increase with the age of regeneration, but these species are commonly found in the secondary forest.

Soil properties change during secondary succession in Imperata grassland area. The effects of secondary succession on soil are strongest in the A-horizon (0–10 cm), where an increase in carbon stock, N, and C/N ratio, and a decrease in bulk density and pH are observed. Soil carbon stocks also increase upon secondary succession from *Imperata* grassland to secondary forest.

Secondary succession in "Imperata"-dominated grassland

Oak and Hickory Forest

A classic example of secondary succession occurs in oak and hickory forests cleared by wildfire. Wildfires will burn most vegetation and kill those animals unable to flee the area. Their nutrients, however, are returned to the ground in the form of ash. Thus, even when areas are devoid of life due to severe fires, the area will soon be ready for new life to take hold. Before the fire, the vegetation was dominated by tall trees with access to the major plant energy resource: sunlight. Their height gave them access to sunlight while also shading the ground and other low-lying species. After the fire, though, these trees are no longer dominant. Thus, the first plants to grow back are usually annual plants followed within a few years by quickly growing and spreading grasses and other pioneer species. Due to, at least in part, changes in the environment brought on by the growth of the grasses and other species, over many years, shrubs will emerge along with small pine, oak, and hickory trees. These organisms are called intermediate species. Eventually, over 150 years, the forest will reach its equilibrium point where species composition is no longer changing and resembles the community before the fire. This equilibrium state is referred to as the climax community, which will remain stable until the next disturbance.

Secondary Succession of an Oak and Hickory Forest

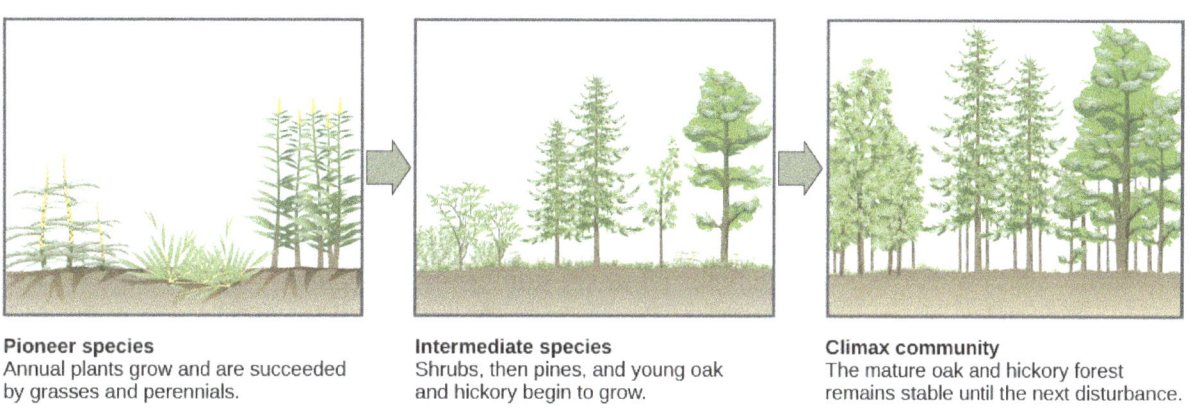

Pioneer species
Annual plants grow and are succeeded by grasses and perennials.

Intermediate species
Shrubs, then pines, and young oak and hickory begin to grow.

Climax community
The mature oak and hickory forest remains stable until the next disturbance.

Post-fire Succession

Soil

Generation of carbonates from burnt plant material following fire disturbance causes an initial increase in soil pH that can affect the rate of secondary succession, as well as what types of organisms will be able to thrive. Soil composition prior to fire disturbance also influences secondary succession, both in rate and type of dominant species growth. For example, high sand concentration was found to increase the chances of primary *Pteridium* over *Imperata* growth in *Imperata* grassland. The byproducts of combustion have been shown to affect secondary succession by soil microorganisms. For example, certain fungal species such as *Trichoderma polysporum* and *Penicillium janthinellum* have a significantly decreased success rate in spore germination within fire-affected areas, reducing their ability to recolonize.

Vegetation

Vegetation structure is affected by fire. In some types of ecosystems this creates a process of renewal. Following a fire, early successional species disperse and establish first. This is then followed by late successional species. Species that are fire intolerant are those that are more flammable and are desolated by fire. More tolerant species are able to survive or disperse in the event of fire. The occurrence of fire leads to the establishment of deadwood and snags in forests. This creates habitat and resources for a variety of species. Fire can act as a seed dispersing stimulant. Many species require fire events to reproduce, disperse, and establish. For example, the knobcone pine ("Pinus attenuata") has closed cones that open for dispersal when exposed to heat caused by forest fires. This particular conifer grows in clusters because of this limited method of seed dispersal. A tough fire resistant outer bark and lack of low branches help the knobcone pine survive fire with minimal damage.

Autogenic Succession

After the succession has begun, the communities itself modify its own environment and thus causing its own replacement.

Example: In a xerarch succession taking place on the rock surface, lichens (pioneer community) produces lichen acids. These lichen acids wither the rock to form the sand. The dead and decayed lichens are added to the sand and form a thin layer of soil. Once the soil is formed, the lichens are immediately replaced by the mosses. Thus the modification of environment by lichen causes its own replacement. The collapse of an aquatic ecosystem due to algal bloom is another example.

Allogenic Succession

Succession that results from factors external to the community is called allogenic succession. In allogenic succession, the principal force of change comes primarily from outside the community. Such external forces may include climate change, changes in temperature and other environmental factors, or other types of massive disturbances. Allogenic succession occurs on a time scale which is in accordance or proportionate with the time scale of the disturbance. Allogenic succession resulting from climate change may occur over thousands of years. Seasonal changes in temperature and light intensity in freshwater lakes are responsible for succession of their phytoplankton communities.

Ecological Pyramid

An ecological pyramid (also trophic pyramid, energy pyramid, or sometimes food pyramid) is a graphical representation designed to show the biomass or bio productivity at each trophic level in a given ecosystem. Biomass is the amount of living or organic matter present in an organism. Biomass pyramids show how much biomass is present in the organisms at each trophic level, while productivity pyramids show the production or turnover in biomass. Ecological pyramids begin with producers on the bottom (such as plants) and proceed through the various trophic levels (such as herbivores that eat plants, then carnivores that eat herbivores, then carnivores that eat those carnivores, and so on). The highest level is the top of the chain. An ecological pyramid of biomass shows the relationship between biomass and trophic level by quantifying the biomass present at each trophic level of an ecological community at a particular time. It is a graphical representation of biomass (total amount of living or organic matter in an ecosystem) present in unit area in different tropic levels. Typical units are grams per meter^2, or calories per meter^2.

General Concepts

Energy flows through the food chain in a predictable way, entering at the base of the food chain, by photosynthesis in primary producers, and then moving up the food chain to higher trophic levels. Because the transfer of energy from one trophic level to the next is inefficient, there is less energy entering higher trophic levels.

It may also be useful and productive to examine how the number and biomass of organisms vary across trophic levels. Both the number and biomass of organisms at each trophic level should be influenced by the amount of energy entering that trophic level. When there is a direct correlation between energy, numbers, and biomass then biomass pyramids and numbers pyramids will result. However, the relationship between energy, biomass, and number can be complicated by the growth form and size of organisms and ecological relationships occurring among trophic levels. Thus, it is possible, and common that biomass pyramids and numbers pyramids do not look like pyramids at all.

Types

There are 3 types of ecological pyramids as described as follows:

- Pyramid of energy
- Pyramid of numbers and
- Pyramid of biomass.

Pyramid of Energy

The pyramid of energy or the energy pyramid describes the overall nature of the ecosystem. During the flow of energy from organism to other, there is considerable loss of energy in the form of heat. The primary producers like the autotrophs there is more amount of energy available. The least energy is available in the tertiary consumers. Thus, shorter food chain has more amount of energy available even at the highest trophic level.

- The energy pyramid always upright and vertical.
- This pyramid shows the flow of energy at different trophic levels.
- It depicts the energy is minimum as the highest trophic level and is maximum at the lowest trophic level.
- At each trophic level, there is successive loss of energy in the form of heat and respiration, etc.

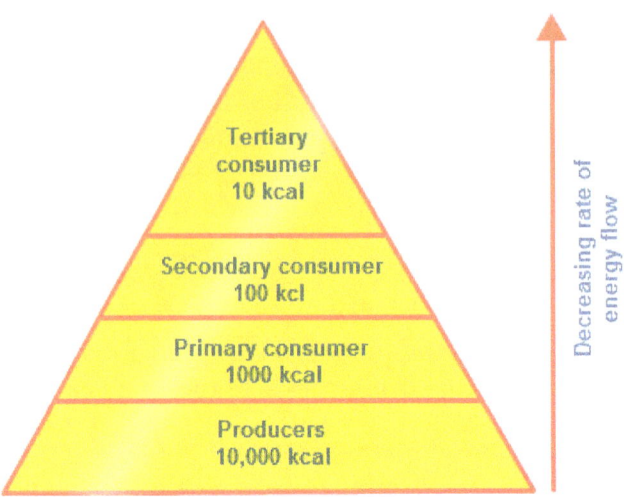

Pyramid of Energy

Pyramid of Numbers

The pyramid of numbers depicts the relationship in terms of the number of producers, herbivores and the carnivores at their successive trophic levels. There is a decrease in the number of individuals from the lower to the higher trophic levels. The number pyramid varies from ecosystem to ecosystem. There are three of pyramid of numbers:

- Upright pyramid of number

- Partly upright pyramid of number and

- Inverted pyramid of number.

Upright Pyramid of Number

This type of pyramid number is found in the aquatic and grassland ecosystem, in these ecosystems there are numerous small autotrophs which support lesser herbivores which in turn support smaller number of carnivores and hence this pyramid is upright.

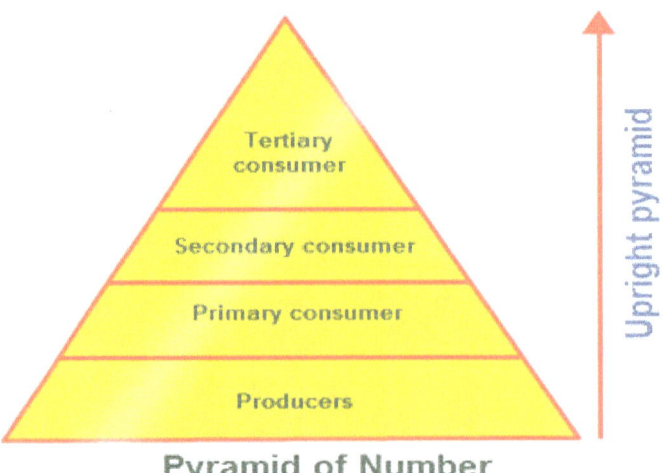

Pyramid of Number

Partly Upright Pyramid of Number

It is seen in the forest ecosystem where the number of producers are lesser in number and support a greater number of herbivores and which in turn support a fewer number of carnivores.

Partly Upright
Pyramid of Number

Inverted Pyramid of Number

This type of ecological pyramid is seen in parasitic food chain where one primary producer supports numerous parasites which support more hyperparasites.

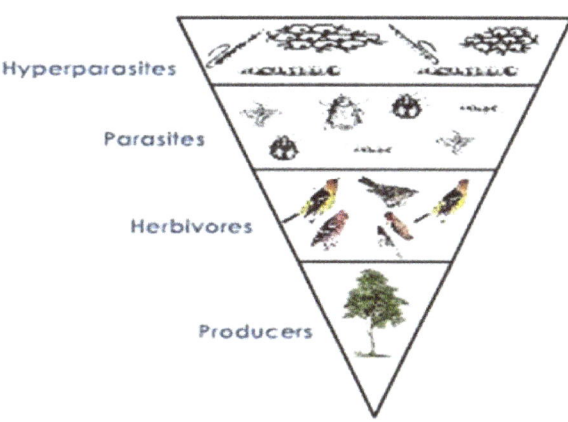

Inverted pyramid of number

Pyramid of Biomass

The pyramid of biomass is more fundamental, they represent the quantitative relationships of the standing crops. In this pyramid there is a gradual decrease in the biomass from the producers to the higher trophic levels. The biomass here the net organisms collected from each feeding level and are then dried and weighed. This dry weight is the biomass and it represents the amount of energy available in the form of organic matter of the organisms. In this pyramid the net dry weight is plotted to that of the producers, herbivores, carnivores, etc.

There are two types of pyramid of biomass, they are:

- Upright pyramid of biomass and
- Inverted pyramid of biomass.

Upright Pyramid of Biomass

This occurs when the larger net biomass of producers support a smaller weight of consumers.

Example: Forest ecosystem.

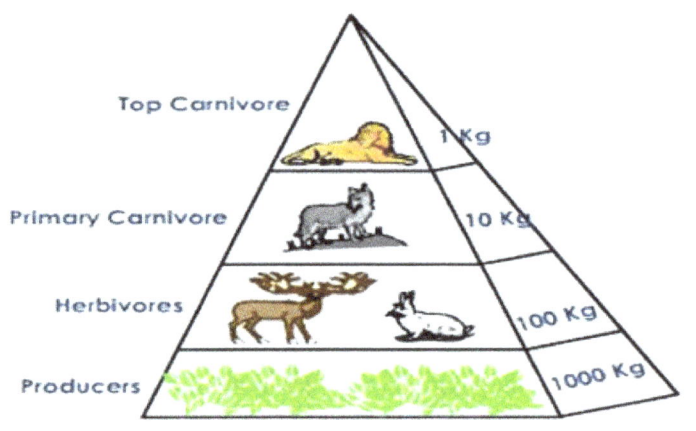

Upright Pyramid of biomass in a Terrestrial Ecosystem

Inverted Pyramid of Biomass

This happens when the smaller weight of producers support consumers of larger weight.

Example: Aquatic ecosystem.

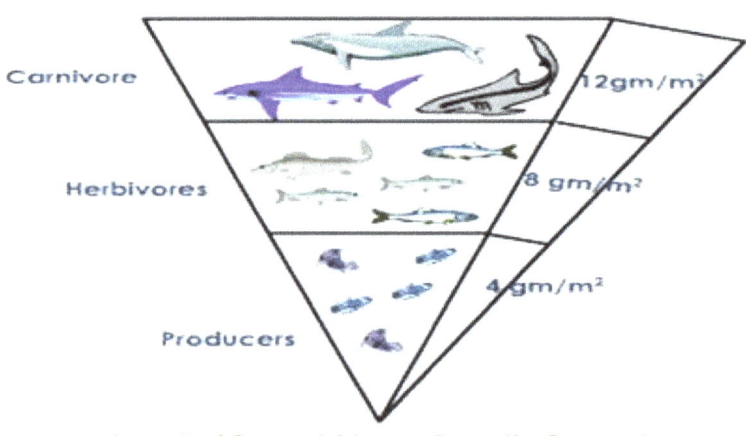

Inverted Pyramid in an Aquatic Ecosystem

Biogeochemical Cycle

We live in a biological ecosystem that shows a constant flow of energy between various organisms. There is an exchange of nutrients, which basically translates to exchange of energy.

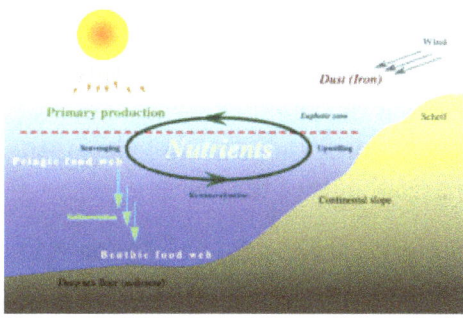

Nutrients ultimately are chemical compounds. So, there is this natural pathway where the living matter is constantly circulated. Nutrients are never lost from an ecosystem. Recycling of these nutrients happens at every stage. This *recycling of the nutrients through different components in an ecosystem is called the nutrient cycle or biogeochemical cycle.*

Here, the chemical elements are always recycled, whereas heat is dissipated. Energy flows but the matter is always recycled. To get a better idea, see the movement of water, which is a chemical compound (H_2O). It moves between different living and non-living forms in various places in the biosphere.

Types of Biogeochemical Cycles

The biogeochemical cycles can be divided into two types, the gaseous biogeochemical cycle and sedimentary biogeochemical cycle based on the reservoir. Each reservoir in a nutrient cycle consists of an abiotic portion and an exchange pool, where there is a rapid exchange that occurs between the biotic and abiotic aspects.

The *gaseous cycles* exist in the atmosphere (air) or Oceans through evaporation. The different gaseous cycles are nitrogen cycle, carbon cycle, oxygen cycle and the water cycle.

The *sedimentary cycles* have the earth's crust as the reservoir pool. These cycles include the chemical components that are more earthbound, such as iron, calcium, sulphur etc. The gaseous cycles move more rapidly when compared to the sedimentary cycles. One of the primary reasons for this could be the large atmospheric reservoir.

> Biogeochemical = Biological Chemical + Geological Process

Thus we can say that with the help of these two processes, the biogeochemical cycles are possible. Some of the major biogeochemical cycles are:

Water Cycle

Water is absorbed from the soil by plants. When transpiration occurs in plants, they release water.

In animals, most of the body is made up of water. Animals also drink water. Also when they perspire, water is released and gets evaporated into the atmosphere.

When animals are eaten by other animals, the water goes from one organism to the other. When animals and plants decompose, they release water, due to the chemical processes that occur. In this way, there is a continuous recycling of water through the various components of the ecosystem.

Carbon Cycle

The carbon cycle is the cycle by which the element carbon moves through our Earth's various systems. It is a fascinating and complex process because living things, atmospheric changes, ocean chemistry, and geologic activity are all part of this cycle.

The Earth is a closed system; that is, it cannot gain or lose materials or resources. Instead, materials must be used and re-used many times over for life to be sustainable. This applies to materials including carbon, oxygen, nitrogen, and water.

Carbon is an essential element for life, as we know it because of its ability to form multiple, stable bonds with other molecules.

This is why nucleotides, amino acids, sugars, and lipids all depend on carbon backbones: carbon provides a stable structure that allows the chemistry of life to happen. Without carbon, none of these molecules could exist and function in the ways that permit the chemistry of life to occur.

Our type of life is referred to as "carbon-based life" by scientists. Some scientists speculate that life on other planets could be based on a different chemistry, but no evidence such life has yet been found.

Originally, carbon was created through nuclear fusion in the hearts of stars. When those stars exploded, materials including carbon were scattered through space. And when the Earth was formed – it was formed of some of those carbon-containing materials.

In time, living things developed as a result of carbon and other atoms reacting in the chemically volatile atmosphere of ancient Earth. When some of these molecules became self-replicating, they began to take carbon from the atmosphere and even from rocks and use it to build materials for life such as sugars, proteins, and lipids.

Early photo synthesizers such as cyanobacteria transformed Earth's atmosphere by turning huge amounts of carbon dioxide gas into oxygen gas. This was not their goal – but rather a happy side effect of these first life forms removing the "C" from CO_2. This left O_2 in the atmosphere – and countless molecules of carbon serving as part of the machinery for life.

In time, other forms of life discovered that the new O_2 in the atmosphere could be used to power a highly efficient method of liberating the energy stored in carbon-based organic molecules like sugars. This allowed these life forms to use the energy plants used to make these molecules for their own metabolism.

Using this process of "cellular respiration," animals and other oxygen-breathers started turning O_2 back into CO_2 – effectively spitting out the carbon atoms once contained in sugars, proteins, and lipids after extracting all of their energy.

This happy balance of plants turning carbon dioxide into living matter while animals reduce it back into gases for the plants to consume has existed for billions of years. In the process, new steps became incorporated – such as the formation of fossil fuels, which occurs when organic matter such as dead plants and animals become trapped underground by geologic processes.

Recently, humans have made some big changes to the Earth's carbon cycle.

By burning huge amounts of fossil fuels and cutting down roughly half of the Earth's forests, humans have decreased the Earth's ability to take carbon out of the atmosphere, while releasing large amounts of carbon into the atmosphere that had been stored in solid form as plant matter and fossil fuels.

This means more carbon dioxide in Earth's atmosphere – which is particularly dangerous since carbon dioxide is a "greenhouse gas" that plays a role in regulating the Earth's temperature and weather patterns.

Function of Carbon Cycle

The carbon cycle, under normal circumstances, works to ensure the stability of variables such as the Earth's atmosphere, the acidity of the ocean, and the availability of carbon for use by living things.

Each of its components is of crucial importance to the health of all living things – especially humans, who rely on many food crops and animals to feed our large population.

Carbon dioxide in the atmosphere prevents the sun's eat from escaping into space, very much like the glass walls of a greenhouse. This isn't always a bad thing – some carbon dioxide in the atmosphere is good for keeping the Earth warm and its temperature stable.

But Earth has experienced catastrophic warming cycles in the past, such as the Permian extinction, which is thought to have been caused by a drastic increase in the atmosphere's level of greenhouse gases.

No one is sure what caused the change that brought about the Permian extinction. Greenhouse gases may have been added to an atmosphere by an asteroid impact, volcanic activity, or even massive forest fires.

Whatever the cause, during this warming episode, temperatures raised drastically. Much of the Earth became desert, and over 90% of all species living at that time went extinct.

This is a good example of what can happen if our planet's essential cycles experience a big change.

Another important variable affected by the carbon cycle is the acidity of the ocean. Carbon dioxide can react with ocean water to form carbonic acid. This has been an important stabilizing force of the carbon cycle over the years, since the chemical equilibrium between carbon dioxide and carbonic acid means that the ocean can absorb or release carbon dioxide as atmospheric levels rise and fall.

However, as you might guess, increasing ocean acidity can mean trouble for sea life – and this might eventually pose a problem for other parts of the carbon system. Many forms of sea life that have shells, for example, can take carbon out of the water to create the calcium carbonate that they make their shells out of. If these species suffer, the ocean may lose some of its ability to remove carbon from the atmosphere.

Lastly, of course, there is the role of living things in the carbon cycle. The activity of plants and animals has been one of the major forces affecting changes to the carbon cycle in the past several billion years. Photo synthesizers have changed Earth's atmosphere and climate drastically by taking huge amounts of carbon out of the atmosphere and turning that carbon into cellular materials.

Those activities created free oxygen and the ozone layer, and generally set the stage for the evolution of animals that obtain their energy by breaking down the organic materials created by photo synthesizers and extracting the energy that the photo synthesizers used to make those molecules.

With one particular species of animals – humans – making big changes, the future of the Earth's carbon cycle is uncertain.

All such cycles in closed systems eventually correct themselves – but sometimes this happens through drastic population reduction of the offending species through starvation.

Steps of Carbon Cycle

Carbon in the Atmosphere

To become part of the carbon cycle, carbon atoms start out in gaseous form. Carbon dioxide gas – CO_2 – can be produced by inorganic processes, or by the metabolisms of living things.

Before Earth had life on it, carbon dioxide gas likely came from volcanic activity and asteroid impacts.

Today, carbon is also released into the atmosphere through the activities of living things, such the exhalations of animals, the actions of decomposer organisms, and the burning of wood and fossil fuels by humans.

However carbon dioxide gets into the atmosphere, CO_2 gas is the starting point of the carbon cycle. The next step is,

Producers Absorb Carbon

"Producers" – organisms that produce food from sunlight, such as plants – absorb carbon dioxide from the atmosphere and use it to build sugars, lipids, proteins, and other essential building blocks of life.

For plants, CO_2 is absorbed through pores in their leaves called "stomata." Carbon dioxide enters the plant through the stomata and is incorporated into containing carbon compounds with the help of energy from sunlight.

Plants and other producer organisms such as cyanobacteria are crucial to life on Earth, because they can turn atmospheric carbon into living matter.

Producers are Eaten

"Consumers" are organisms that eat other living things. Animals are the most visible type of consumer in our ecosystems, though many types of microbes also fall into this category.

Consumers incorporate the carbon compounds from plants and other food sources when they eat them.

They use some of these carbon compounds from food to build their own bodies – but much of the food they eat is broken down to release energy, in a process that is almost the reverse of what producers do.

While producers use energy from sunlight to make bonds between carbon atoms – animals break these bonds to release the energy they contain, ultimately turning sugars, lipids, and other carbon compounds into single-carbon units. These are ultimately released into the atmosphere in the form of CO_2.

This process of "cellular respiration" – where oxygen gas is inhaled and carbon dioxide is exhaled – is a major source of carbon release back into the atmosphere.

But it's not always the last step of the carbon cycle. What about the carbon compounds that don't get eaten, or broken down by animals?

Decomposers Release Carbon

Plants and animals that die without being eaten by other animals are broken down by other organisms, called "decomposers."

Decomposers include many bacteria and some fungi. They usually only break down matter that is already dead, rather than catching and eating a living animal or plant.

Just like animals, decomposers break down the chemical bonds in their food molecules. They create many chemical produces, including in some cases CO_2.

Human Combustion

Humans are the only animals we know of who can create fire on purpose. And we set fire to things *a lot*.

Our cars are driven by burning fossil fuels – oil and gasoline, which are made of dead plant and animal material that spent millions of years buried deep in the Earth.

Many of our electrical power plants are powered by burning fossil fuels as well, including coal, which is another form of dead plant matter that was buried underground and transformed by geologic heat.

Lastly, humans also burn a lot of wood. We no longer burn wood to power our machines as we did in the 19th century, but now we often burn forests in order to clear land for agriculture, mining, and other purposes. About half of Earth's forests have been burned or otherwise destroyed by human activity to date.

The scientific community has raised alarms that by making significant changes to the Earth's carbon cycle, we may end up changing our climate or other important aspects of the ecosystem we rely upon to survive.

As a result, many scientists advocate to decrease the amount of carbon burned by humans by reducing car trips and electricity consumption, and investing in non-burning sources of energy such as solar power and wind power.

Examples of the Carbon Cycle

The carbon cycle consists of many parallel systems, which can either absorb or release carbon. Together, these systems work to keep Earth's carbon cycle – and subsequently its climate and biosphere – relatively stable.

Here are some examples of parts of Earth's ecosystems that can absorb carbon, turn carbon into living matter, or release carbon back into the atmosphere.

Atmosphere

One major repository of carbon is the carbon dioxide in the Earth's atmosphere. Carbon forms a stable, gaseous molecule in combination with two atoms of oxygen.

In nature, this gas is released by volcanic activity, and by the respiration of animals who affix carbon molecules from the food they eat to molecules of oxygen before exhaling it.

Humans also release carbon dioxide into the atmosphere by burning organic matter such as wood and fossil fuels.

Carbon dioxide can be removed from the atmosphere by plants, which take the atmospheric carbon and turn it into sugars, proteins, lipids, and other essential molecules for life.

It can also be removed from the atmosphere by absorption into the ocean, whose water molecules can bond with carbon dioxide to form carbonic acid.

In recent years, scientists have raised concerns that by cutting down about half of Earth's forests, humans may be decreasing the Earth's ability to remove carbon dioxide from the atmosphere at the same time they're adding new sources by burning wood and fossil fuels.

Many organizations that hope to fight man-made climate change now plant trees which can remove carbon dioxide from the atmosphere, as well as advocating for alternative energy sources and less burning of fossil fuels such as gasoline, oil, and coal.

Lithosphere

The Earth's crust – called the "lithosphere" from the Greek word "litho" for "stone" and "sphere" for globe – can also release carbon dioxide into Earth's atmosphere. This gas can be created by chemical reactions in the Earth's crust and mantle.

Volcanic activity can result in natural releases of carbon dioxide. Some scientists believe that widespread volcanic activity may be to blame for the warming of the Earth that caused the Permian extinction.

While the Earth's crust can add carbon to the atmosphere, it can also remove it. Movements of the Earth's crust can bury carbon-containing chemicals such as dead plants and animals deep underground, where their carbon cannot escape back into the atmosphere.

Over millions of years, these underground reservoirs of organic matter liquefy and become coal, oil, and gasoline. In recent years, humans have begun releasing much of this sequestered carbon back into the atmosphere by burning these materials to power cars, power plants, and other human equipment.

Biosphere

Among living things, some remove carbon from the atmosphere, while others release it back. The most noticeable participants in this system are plants and animals.

Plants remove carbon from the atmosphere. They don't do this as a charitable act; atmospheric carbon is actually the "food" which plants use to make sugars, proteins, lipids, and other essential molecules for life.

Plants use the energy of sunlight, harvested through photosynthesis, to build these organic compounds out of carbon dioxide and other trace elements. Indeed, the term "photosynthesis" comes from the Greek words "photo" for "light" and "synthesis" for "to put together."

Here lies arguably the most important part of the carbon cycle: all life is made of carbon. Without plants or other organisms that could turn inorganic carbon compounds like carbon dioxide into organic compounds, life could not exist.

Indeed, none of the building materials for our cells, from our DNA to our cell membranes, could exist without this ability of photo synthesizers to turn carbon dioxide into life!

In a gracefully balanced set of chemical reactions, animals eat plants (and other animals), and take these synthesized molecules apart again. Animals get their fuel from the chemical energy plants have stored in the bonds between carbon atoms and other atoms during photosynthesis.

In order to do that, animal cells dissemble complex molecules such as sugars, fats, and proteins all the way down to single-carbon units – molecules of carbon dioxide, which are produced by reacting carbon-containing food molecules with oxygen from the air.

In this way, most of the carbon eaten by animals ends up back where it started before it was absorbed by a plant – as part of a carbon dioxide molecule in the atmosphere.

Carbon in plants and animals that is not consumed by other animals can be broken down by other living things until it becomes carbon dioxide again, or can be sequestered deep in the Earth as fossil fuels.

Oceans

The Earth's oceans have the ability to both absorb and release carbon dioxide. When carbon dioxide from the atmosphere comes into contact with ocean water, it can react with the water molecules to form carbonic acid – a dissolved liquid form of carbon.

Like most chemical reactions, the rate of this reaction is determined by the equilibrium between the products and the reactants.

When there is more carbonic acid in the ocean compared to carbon dioxide in the atmosphere, some carbonic acid may be released into the atmosphere as carbon dioxide.

On the other hand, when there is more carbon dioxide in the atmosphere, more carbon dioxide will be converted to carbonic acid, and ocean acidity levels will rise.

Some scientists have raised concerns that acidity is rising in some parts of the ocean, possibly as a result of increased carbon dioxide in the atmosphere due to human activity.

Although these changes in ocean acidity may sound small by human standards, many types of sea life depend on chemical reactions that need a highly specific acidity level to survive.

As a result, there is concern that increasing ocean acidity due to carbonic acid may contribute to the die-offs of some marine ecosystems, and even to extinctions of marine species.

Nitrogen Cycle

The nitrogen cycle is the biogeochemical cycle that describes the transformations of nitrogen and nitrogen-containing compounds in nature.

The basic Earth's atmosphere is about 78 percent nitrogen, making it the largest pool of nitrogen. Nitrogen is essential for many biological processes; it is in all amino acids, is incorporated into proteins, and is present in the bases that make up nucleic acids, such as DNA and RNA. In plants, much of the nitrogen is used in chlorophyll molecules, which are essential for photosynthesis and further growth.

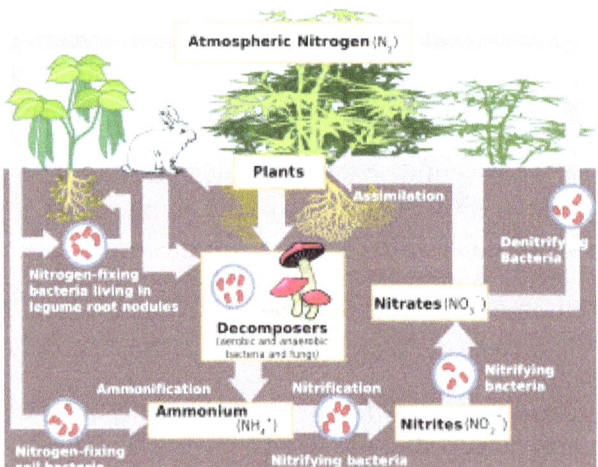

Schematic representation of the flow of nitrogenthrough the environment. The importance of bacteria in the cycle is immediately recognized as being a key element in the cycle, providing different forms of nitrogen compounds assimilable by higher organisms.

The nitrogen cycle reveals the harmonious coordination between different biotic and abiotic elements. Processing, or fixation, is necessary to convert gaseous nitrogen into forms usable by living organisms. Some fixation occurs in lightning strikes, but most fixation is done by free-living or symbiotic bacteria. These bacteria have the nitrogenase enzyme that combines gaseous nitrogen with hydrogen to produce ammonia, which is then further converted by the bacteria to make its own organic compounds. Some nitrogen-fixing bacteria, such as *Rhizobium*, live in the root nodules of legumes (such as peas or beans). Here they form a mutualistic relationship with the plant, producing ammonia in exchange for carbohydrates. Nutrient-poor soils can be planted with legumes to enrich them with nitrogen. A few other plants can form such symbioses.

Other plants get nitrogen from the soil by absorption at their roots in the form of either nitrate ions or ammonium ions. All nitrogen obtained by animals can be traced to the eating of plants at some stage of the food chain.

Ammonia

The source of ammonia is the decomposition of dead organic matter by bacteriacalled *decomposers*, which produce ammonium ions (NH_4^+). In well-oxygenated soil, these ions are then oxygenated first by nitrifying bacteria into nitrite (NO_2^-) and then into nitrate (NO_3^-). This two-step conversion of ammonium into nitrate is called *nitrification*.

Ammonia is highly toxic to fish life and the water discharge level of ammonia from wastewater treatment plants must often be closely monitored. To prevent loss of fish, nitrification prior to discharge is often desirable. Land application can be an attractive alternative to the mechanical aeration needed for nitrification.

Ammonium ions readily bind to soils, especially to humic substances and clays. Nitrate and nitrite ions, due to their negative electric charge, bind less readily since there are less positively charged ion-exchange sites (mostly humic substances) in soil than negative. After rain or irrigation, *leaching* (the removal of soluble ions, such as nitrate and nitrite) into groundwater can occur. Elevated nitrate in groundwater is a concern for drinking water use because nitrate can interfere with blood-oxygen levels in infants and cause methemoglobinemia or blue-baby syndrome. Where groundwater recharges stream flow, nitrate-enriched groundwater can contribute to eutrophication, a process leading to high algal and blue-green bacterial populations and the death of aquatic life due to excessive demand for oxygen. While not directly toxic to fish life as is ammonia, nitrate can have indirect effects on fish if it contributes to this eutrophication. Nitrogen has contributed to severe eutrophication problems in some water bodies. As of 2006, the application of nitrogen fertilizer is being increasingly controlled in the United Kingdom and the United States. This is occurring along the same lines as control of phosphorus fertilizer, restriction of which is normally considered essential to the recovery of eutrophied waterbodies.

During anaerobic (low oxygen) conditions, *denitrification* by bacteria occurs. This results in nitrates being converted to nitrogen gas and returned to the atmosphere.

Processes of the Nitrogen Cycle

Nitrogen Fixation

There are three main ways to convert N_2 (atmospheric nitrogen gas) into more chemically reactive forms:

- *Biological fixation*; some symbiotic bacteria (most often associated with leguminous plants) and some free-living bacteria are able to fix nitrogen and assimilate it as organic nitrogen. An example of a mutualistic nitrogen fixing bacteria is the *Rhizobium* bacteria, which lives in plant root nodes. As well, there are free living bacteria, typically in the soil, such as the *Azotobacter*, that are responsible for nitrogen fixation.

- *Industrial N-fixation*; in the Haber-Bosch process, N_2 is converted together with hydrogen gas (H_2) into ammonia (NH_3) fertilizer.

- *Combustion of fossil fuels*; automobile engines and thermal power plants, which release NOx.

Additionally, the formation of NO from N_2 and O_2 due to photons and lightning, is important for atmospheric chemistry, but not for terrestrial or aquatic nitrogen turnover.

As a result of extensive cultivation of legumes (particularly soy, alfalfa, and clover), use of the Haber-Bosch process in the creation of chemical fertilizers, and pollution emitted by vehicles and industrial plants, human beings are estimated to have more than doubled the annual transfer of nitrogen into a biologically available form. This has occurred to the detriment of aquatic and wetland habitats through eutrophication.

Nitrification

Nitrification is the biological oxidation of ammonia with oxygen into nitrite followed by the oxidation of these nitrites into nitrates. Nitrification is an important step in the nitrogen cycle in soil. This process was discovered by the Russian microbiologist, Sergei Winogradsky.

The oxidation of ammonia into nitrite, and the subsequent oxidation to nitrate is performed by two different bacteria. The first step is done by bacteria of (among others) the genus *Nitroso-monas* and *Nitrosococcus*. The second step (oxidation of nitrite into nitrate) is (mainly) done by bacteria of the genus *Nitrobacter*. All organisms are autotrophs, which means that they take carbon dioxide as their carbon source for growth. In most environments, both organisms are found together, yielding nitrate as the final product. It is possible however to design systems in which selectively nitrite is formed (the *Sharon process*).

Nitrification also plays an important role in the removal of nitrogen from municipal wastewater. The conventional removal is nitrification, followed by denitrification. The cost of this process resides mainly in aeration (bringing oxygen in the reactor) and the addition of an extra organic energy source (e.g. methanol) for the denitrification.

Together with ammonification, nitrification forms a mineralization process which refers to the complete decomposition of organic material, with the release of available nitrogen compounds. This replenishes the nitrogen cycle. Nitrification is a process of nitrogen compound oxidation (effectively, loss of electrons from the nitrogen atom to the oxygen atoms).

Assimilation

In plants that have a mutualisic relationship with Rhizobium, some nitrogen is assimilated in the form of ammonium ions from the nodules. All plants, however, can absorb nitrate from the soil via their root hairs. These are then reduced to nitrate ions and then ammonium ions for incorporation into amino acids, and hence protein, which forms part of the plants or animals that they eat.

Ammonification

Nitrates are the form of nitrogen most commonly assimilated by plant species, which, in turn are consumed by heterotrophs for use in compounds such as amino and nucleic acids. The remains of heterotrophs will then be decomposed into nutrient rich organic material and bacteria or in some cases, fungi will convert the nitrates within the remains back into ammonia.

Denitrification

Denitrification is the process of reducing nitrate, a form of nitrogen available for consumption by many groups of organisms, into gaseous nitrogen, which is far less accessible to life forms, but makes up the bulk of our atmosphere. It can be thought of as the opposite of nitrogen fixation, which converts gaseous nitrogen into more biologically useful forms. The process is performed by heterotrophic bacteria (such as *Pseudomonas fluorescens*) from all main proteolitic groups. Denitrification and nitrification are parts of the nitrogen cycle.

Denitrification takes place under special conditions in both terrestrial and marine ecosystems. In general, it occurs when oxygen (which is a more favorable electron acceptor) is depleted, and bacteria turn to nitrate in order to respire organic matter. Because our atmosphere is rich with oxygen, denitrification only takes place in some soils and groundwater, wetlands, poorly ventilated corners of the ocean, and in seafloor sediments.

Denitrification proceeds through some combination of the following steps:

nitrate → nitrite → nitric oxide → nitrous oxide → dinitrogen gas

Or expressed as a redox reaction:

$$2NO_3^- + 10e^- + 12H^+ \rightarrow N_2 + 6H_2O$$

Denitrification is the second step in the nitrification-denitrification process: the conventional way to remove nitrogen from sewage and municipal wastewater.

Direct reduction from nitrate to ammonium (a process known as DNRA) is also possible for organisms that have the nrf-gene.

In some wastewater treatment plants, a small amount of methanol is added to the wastewater to provide a carbon source for the denitrification bacteria.

Human Influences on the Nitrogen Cycle

Humans have contributed significantly to the nitrogen cycle by artificial nitrogen fertilization (primarily through the Haber Process; using energy from fossil fuels to convert N_2 to ammonia gas (NH_3)); and planting of nitrogen fixing crops. In addition, humans have significantly contributed to the transfer of nitrogen trace gases from Earth to the atmosphere.

N_2O has risen in the atmosphere as a result of agricultural fertilization, biomass burning, cattle and feedlots, and other industrial sources. N_2 has deleterious effects in the stratosphere, where it breaks down and acts as a catalyst in the destruction of atmospheric ozone.

NH_3 in the atmosphere has tripled as the result of human activities. It is a reactant in the atmosphere, where it acts as an aerosol, decreasing air quality and clinging on to water droplets, eventually resulting in acid rain.

Fossil fuel combustion has contributed to a six- or seven-fold increase in NOx flux to the Earth's atmosphere. NO actively alters atmospheric chemistry, and is a precursor of tropospheric (lower atmosphere) ozone production, which contributes to smog and acid rain, increasing nitrogen inputs to ecosystems.

Ecosystem processes can increase with nitrogen fertilization, but anthropogenic input can also result in nitrogen saturation, which weakens productivity and can kill plants. Decreases in biodiversity can also result if higher nitrogen availability increases nitrogen-demanding grasses, causing a degradation of nitrogen-poor, species-diverse heathlands.

Rock Cycle

Like most Earth materials, rocks are created and destroyed in cycles. The rock cycle is a model that describes the formation, breakdown, and reformation of a rock as a result of sedimentary, igneous, and metamorphic processes. All rocks are made up of minerals. A mineral is defined as a naturally occurring, crystalline solid of definite chemical composition and a characteristic crystal structure. A rock is any naturally formed, nonliving, firm, and coherent aggregate mass of solid matter that constitutes part of a planet.

Igneous rocks- form in two very different environments. All igneous rocks start out as melted rock, (magma) and then crystallize, or freeze. Bowen's Reaction Series is a proposed sequence of mineral crystallization from basaltic magma, based on experimental evidence. Volcanic processes form extrusive igneous rocks. Extrusive rocks cool quickly on or very near the surface of the earth. Fast cooling makes crystals too small to see without some kind of magnifier. Basalt is dark rock, gray or black on a freshly broken surface, and weathers brown or red, because it contains lots of dark-colored minerals.

Some basalt contains light-colored crystals. Dacite and andesite are medium in color, and contains medium amounts of dark minerals. Rhyolite is the lightest colored volcanic rock. Rhyolite contains very few dark minerals, but sometimes, rhyolite cools so fast that it quenches and forms volcanic glass instead of crystallizing. Volcanic glass looks dark because of the way light passes through it. Obsidian is volcanic glass. Rhyolite is the most common source of volcanic ash and pumice in Idaho. Intrusive igneous rocks cool in plutons (Pluto was the Roman god of the Underworld.) deep below the surface of the Earth. Slow cooling allows the growth of large crystals. Crystals in intrusive rocks are visible without magnification. Granite has the same minerals as rhyolite, but in much larger crystals. Diorite is the intrusive version of andesite, granodiorite is the intrusive version of dacite, and gabbro is the intrusive version of basalt.

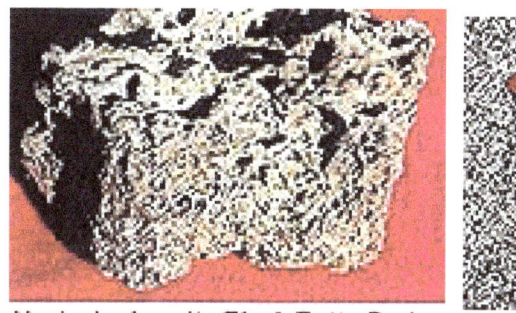

Vesicular basalt - Black Butte Crater

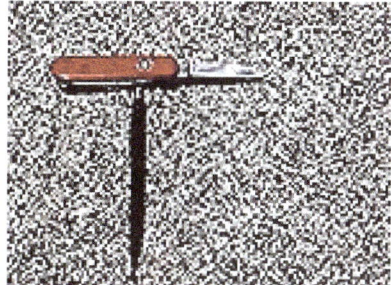

Granodiorite from Soldier Mountains

Rock Cycle

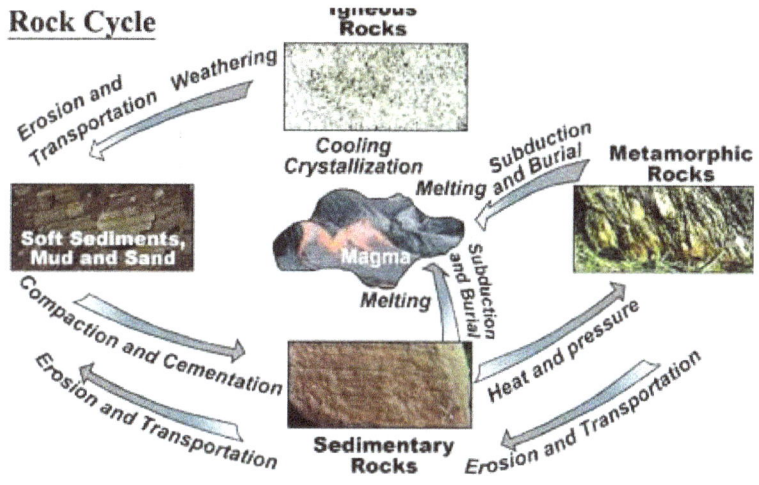

Metamorphic Rock- Metamorphic rocks form when sedimentary, igneous, or other metamorphic rocks are subjected to heat and pressure from burial or contact with intrusive or extrusive igneous rocks. ("Meta" means change, and "morph" means form.) Heat and pressure from burial cause molecules of flat minerals like mica to line up perpendicular to the direction of greatest compression. Deep burial means higher pressure and hotter temperatures, and very high temperature and pressures cause the formation of new minerals, and mineral grains. Low-grade metamorphic rocks like slate and phyllite break in flat pieces, and have a sheen on the surface. Schist is shiny, and many schists contain garnets, staurolites or other mineral crystals that have grown within the rock. Gneiss is a foliated metamorphic rock. Layers of dark and light minerals stripe the rock, and sometimes it is possible to see how the direction of pressure deep in the Earth changed as the minerals formed. The change in direction forms eye-shaped pods of minerals, called augens ("augen" is German for "eye.") Quartzite is another important metamorphic rock in Idaho. Quartzite is metamorphosed sandstone. Some Idaho quartzite is so pure that it can be used to make computer chips. The most common contact metamorphic rock in Idaho is marble. The Portneuf Gap area provides good examples of Idaho marble. Marble forms when limestone is intruded by a pluton which heats the limestone.

Sedimentary Rock- Sedimentary rocks are those rocks made up of pieces of other rocks. We call the pieces of rock "clasts" (*Clast* means "broken piece"). A clast is a piece of rock broken off of another rock. Clasts of rock are eroded from larger rocks, transported (moved) by wind or water and deposited in a basin.After some period of time, the clasts are lithified. The sedimentary rocks we see today were once gravel, sand, silt, mud, or living things. We decide what to name sedimentary rocks based on the size of the clasts that make up the rock. For most sedimentary rocks, this is easy. Sandstone is made of sand, siltstone is made of silt, mudstone is made of mud and so on. Even volcanic ash can become sedimentary rock! The only hard ones to remember are conglomerate and breccia. Conglomerates are made up of rounded, gravel-size particles (To a geologist, gravel is anything from 2mm to 4 meters in diameter), and breccia is made up of angular, sharp-edged, gravel-sized clasts. Limestone and chert are classified as sedimentary rocks, but most limestone and chert are grown by living organisms rather than broken from other rocks. Some limestones have fossils, but most limestones and cherts have recrystallized, and the remains of the creatures that made them are no longer visible.

Mudstone near Preston Idaho

Oxygen Cycle

The oxygen cycle portrays how the flow of oxygen occurs through the several parts of our vast ecosystem. Oxygen is found in several parts of the ecosystem, from the air we breathe (Atmosphere), the water bodies on the planet (Hydrosphere), inside all the biological beings (Biosphere) and inside the earth's crust (Lithosphere).

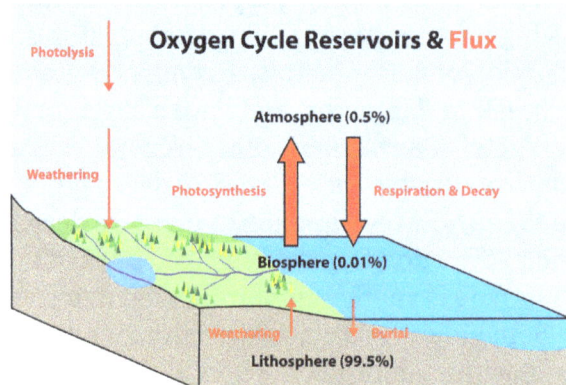

Oxygen Cycle Steps:

- Atmosphere:
  Only a small percentage of the world's oxygen is present in the atmosphere, only about 0.35 %. This exchange of gaseous oxygen happens through Photolysis.

  o Photolysis: This is the process by which molecules like atmospheric water and nitrous oxide are broken down by the ultraviolet radiation coming from the sun and release free oxygen.

- Biosphere:

  The exchange of oxygen between the living beings on the planet, between the animal kingdom and the plant kingdom. The exchange of oxygen in the biosphere is codependent on the Carbon cycle and hydrogen cycle as well. It mainly occurs through 2 processes.

  o Photosynthesis:

     The process by which plants make energy by taking in carbon dioxide from the atmosphere and give out oxygen.

  o Respiration:

     The process by which animals and humans take in oxygen from the atmosphere and use it to break down carbohydrates and give out carbon dioxide.

- Lithosphere:

  The part of the planet containing the most of the oxygen content through biomass, organic content and mineral deposits. These deposits are formed when free radical elements were exposed to free oxygen and over time they form silicates and oxides. This trapped oxygen is released back due to several weathering processes. Also, animals and plants draw nutrient materials from the from the lithosphere and free some of the trapped oxygen.

- Hydrosphere:

  Oxygen dissolved in water is responsible for the sustenance of the aquatic ecosystem present beneath the surface. The hydrosphere is 33% oxygen by volume present mainly as a component of water molecules with dissolved molecules including carbonic acids and free oxygen.

Phosphorus Cycle

The phosphorus cycle differs from the other major biogeochemical cycles in that it does not include a gas phase; although small amounts of phosphoric acid (H_3PO_4) may make their way into the atmosphere, contributing—in some cases—to acid rain. The water, carbon, nitrogen and sulfur cycles all include at least one phase in which the element is in its gaseous state. Very little phosphorus circulates in the atmosphere because at Earth's normal temperatures and pressures, phosphorus and its various compounds are not gases. The largest reservoir of phosphorus is in sedimentary rock.

It is in these rocks where the phosphorus cycle begins. When it rains, phosphates are removed from the rocks (via weathering) and are distributed throughout both soils and water. Plants take up the phosphate ions from the soil. The phosphate then moves from plants to animals when herbivores eat plants and carnivores eat plants or herbivores. The phosphates absorbed by animal tissue through consumption eventually returns to the soil through the excretion of urine and feces, as well as from the final decomposition of plants and animals after death.

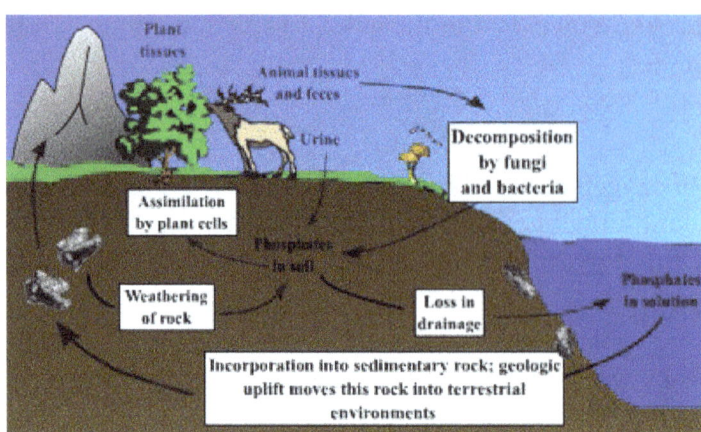

The same process occurs within the aquatic ecosystem. Phosphorus is not highly soluble, binding tightly to molecules in soil; therefore it mostly reaches waters by traveling with runoff soil particles. Phosphates also enter waterways through fertilizer runoff, sewage seepage, natural mineral deposits, and wastes from other industrial processes. These phosphates tend to settle on ocean floors and lake bottoms. As sediments are stirred up, phosphates may reenter the phosphorus cycle, but they are more commonly made available to aquatic organisms by being exposed through erosion. Water plants take up the waterborne phosphate which then travels up through successive stages of the aquatic food chain.

While obviously beneficial for many biological processes, in surface waters an excessive concentration of phosphorus is considered a pollutant. Phosphate stimulates the growth of plankton and plants, favoring weedy species over others. Excess growth of these plants tend to consume large amounts of dissolved oxygen, potentially suffocating fish and other marine animals, while also blocking available sunlight to bottom dwelling species. This is known as eutrophication.

Humans can alter the phosphorus cycle in many ways, including in the cutting of tropical rain forests and through the use of agricultural fertilizers. Rainforest ecosystems are supported primarily through the recycling of nutrients, with little or no nutrient reserves in their soils. As the forest is cut and/or burned, nutrients originally stored in plants and rocks are quickly washed away by heavy rains, causing the land to become unproductive. Agricultural runoff provides much of the phosphate found in waterways. Crops often cannot absorb all of the fertilizer in the soils, causing excess fertilizer runoff and increasing phosphate levels in rivers and other bodies of water. At one timSSe the use of laundry detergents contributed to significant concentrations of phosphates in rivers, lakes, and streams, but most detergents no longer include phosphorus as an ingredient.

Sulphur Cycle

Only a few organisms meet their sulphur requirements in such forms as amino acid and cystein. The source of biologically significant sulphur is inorganic sulphate.

The reservoir of sulphur lies in the soil and sedimentary rocks. The atmosphere is a minor reservoir formed by fuel combustion.

The centre wheel of the sulphur cycle (Figure below) rotates round the activity of a group of specialized micro-organisms which function as a relay team, each carrying out a particular chemical oxidation or reduction.

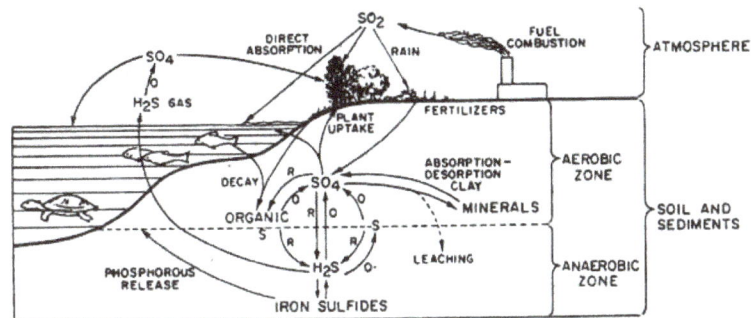

Fig. 3.17.   Showing sulphur cycle.   The central 'wheel-like' diagram is indicative of the oxidation (O) and reduction (R) that bring about the SO_4 pool and the reservoir iron-sulphide pool extended in soils and sediments.

The sedimentary aspect of the cycle involves the precipitation of sulphur in the presence of iron in anaerobic condition. Ferrous sulphide is unsoluble in neutral or alkaline water and as a result the sulphur has the potential for being bound up under these conditions to the limits of the amount of iron present.

The biologically incorporated sulphur is mineralized by bacteria and fungi in ordinary decomposition. Some such sulphur is also reduced directly to sulphides including hydrogen sulphide by bacteria specially the Escherichia and Proteus.

Inorganic sulphate (SO_4) is the source of elemental sulphur in the ecosystems. Under anaerobic condition the sulphate is reduced to elemental sulphur or to hydrogen sulphide by bacteria under the genus Desulphovibrio, Escherichia and Aerobactor. The presence of a large amount of hydrogen sulphide occurring in the anaerobic or deeper portion of aquatic ecosystem is inimical to animal life. The H_2S rises to shallow sediments and is acted upon by other organisms.

Colourless sulphur bacteria such as species of Beggiatoa oxidize hydrogen sulphide to elemental sulphur. Species of Thiobacillus oxidize elemental sulphur to sulphate and other species of Thiobacillus oxidize sulphide to sulphur.

At the global level the regulation of sulphur cycle is dependent upon the interaction of geochemical and meteorological processes (erosion, sedimentation, leaching, rain absorption), and biological processes (production and decomposition). The interdependence of air, soil and water also aids in the regulation.

References

- Abiogenesis, science: britannica.com, Retrieved 22 June 2018

- Cook, W.M.; Yao, J.; Forster, B.L.; Holt, R.D.; Patricks, L.B. (2005). "Secondary succession in an experimentally fragmented landscape: Community pattern across space and time". Ecology. 86: 1267–1279. doi:10.1890/04-0320

- What-is-biogenesis: compellingtruth.org, Retrieved 12 April 2018

- Species-diversity-2992: safeopedia.com, Retrieved 25 June 2018

- Primary-succession: biologydictionary.net, Retrieved 15 May 2018

- Burczyk, Jaroslaw; Adams, W. T.; Shimizu, Jarbas Y. (3 October 1996). "Mating patterns and pollen dispersal in a natural knobcone pine (Pinus attenuata Lemmon.) stand". Heredity. 77: 251, 260. doi:10.1038/sj.hdy.6880410. Retrieved 27 May 2016

- Different-types-of-ecological-succession: easybiologyclass.com, Retrieved 26 April 2018

- Ecological-succession-in-biotic-community-4698: biologydiscussion.com, Retrieved 14 July 2018

- Biogeochemical-cycles: byjus.com, Retrieved 14 April 2018

- Nitrogen-cycle: newworldencyclopedia.org, Retrieved 30 June 2018

The Lithosphere

A lithosphere is the outermost shell of a planet or a natural satellite. In case of the Earth, the lithosphere includes the crust and the upper mantle. The topics elaborated in this chapter on rock cycle, composition of rocks, plate tectonics, mountain building, weathering, etc. address the crucial aspects of the lithosphere and the various processes affecting it.

Rock Cycle

The rock cycle by definition is a natural process by which sedimentary, igneous, and metamorphic rocks are created, changed from one type to another, and destroyed.

3 Types of Rocks

To start to understand the rock cycle we must first understand the three primary types of rocks: sedimentary, igneous, and metamorphic. These rocks are differentiated by their physical properties, chemistry, biology (fossils), but mostly by their origin.

- Sedimentary rocks are rocks formed from the compression of sediments, dirt, or sand we see on the surface of Earth today. As you bury sediment deeper and deeper into the crust, temperatures and pressures increase to the point that the individual grains are cemented together or lithified. Sediments can be either abiogenic or biogenically sourced. Sedimentary rocks often contain fossils from marine organisms or are entirely made up of fossils in the case of many carbonates around the world.

- Igneous rocks are formed from cooling magma deep in Earth's crust or mantle. This cooling magma crystalizes to form rocks like the granite in your house. A rock that cools within Earth's crust will cool very slowly and form larger crystals and is called an intrusive igneous rock. Magma that is ejected to the surface of Earth a volcanic eruption or at a spreading center cools very quickly, contains small crystals typically and is called an extrusive igneous rock.

- Metamorphic rocks are a formed from the partial melting of previously existing material, either sedimentary, igneous, or older metamorphic rocks. Metamorphic rocks are dependent on the degree of melting, where complete melting "resets" the rock to magma and will then form igneous rocks when cooled.

The three types of rocks above can each form one another by melting or erosion and subsequent lithification. This process helps to bring nutrients from deep in Earth's mantle to the surface. This continual recycling of nutrients and elements helps to sustain life on Earth and maintain its biogeochemical processes.

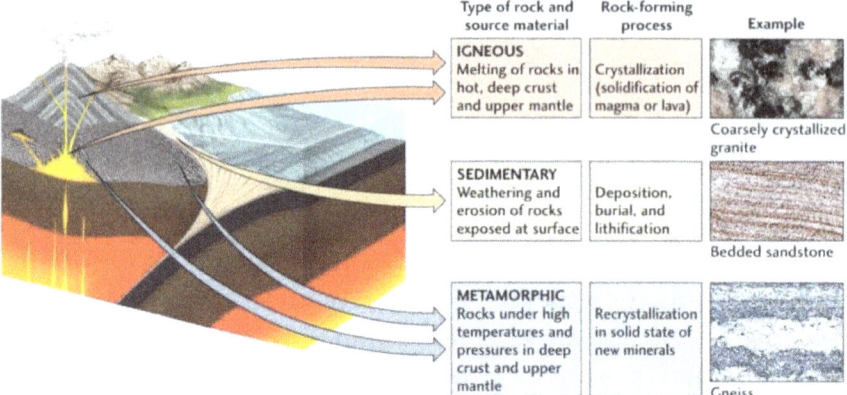

The 3 types of rocks

The Rock Cycle

The rock cycle acts to recycle rocks and the minerals that make up rocks. The best way to explain the cycle is with the rock cycle diagram below and with examples of how each rock type can transform into another rock type.

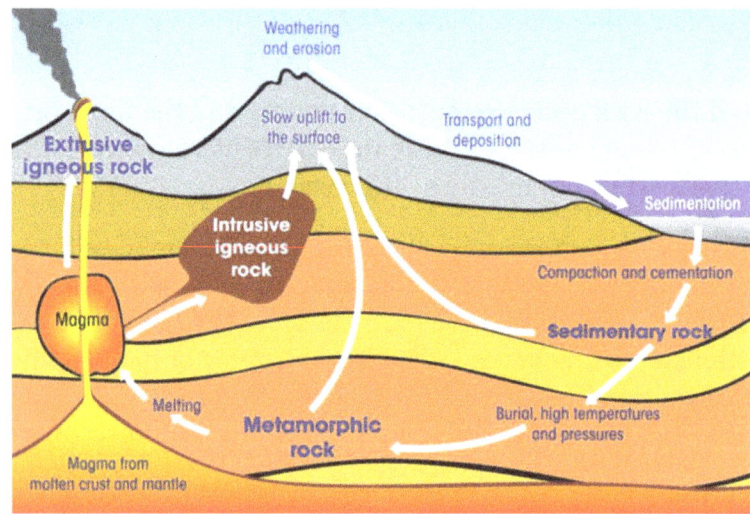

The Rock Cycle

Igneous or metamorphic rocks to sedimentary rocks - As igneous rocks weather and erode from mountains they are transported by water and wind down slope to sedimentary basins such as lakes or oceans. This process results in an accumulation of sediment that continually buries deeper sediment. Imagine how much sediment the Mississippi River transports from the continent into the Gulf of Mexico. This sediment continually piles on top of one another. Eventually the sediment originally from eroded igneous rocks gets buried deep enough that the sediments begin to lithify. High temperatures and pressures compact the sediment enough to expel water and cement the grains into a sedimentary rock. The same process applies for metamorphic rocks, which weather and erode to form sediments that are then lithified.

Igneous or sedimentary rocks to metamorphic rocks - Igneous rocks can form metamorphic rocks by either local contact metamorphism or regional metamorphism. Imagine a granite that finds itself all of a sudden in contact with magma from a nearby volcanic eruption. This contact with extremely hot magma will alter the granite chemically and physically to form gneiss, the metamorphic form of granite.

Granite
igneous
rock

Gneiss
metamorphic
rock

Comparison of granite versus gneiss

Another way to form gneiss from granite is to have high stress or pressure from regional plate tectonics. For example the India plate colliding with the Eurasian plate, part of global plate tectonics. The collision provides enough stress to transform igneous rocks to metamorphic rocks. Lastly, you can bury a rock to depths deep enough to partially melt but not deep enough to completely melt. A similar process takes place with transforming sedimentary rocks into metamorphic rocks. Local heating or regional stress and temperature can be enough to change a quartz sandstone into a metamorphic quartzite.

Metamorphic or sedimentary rocks to igneous rocks - The process of forming igneous rocks from metamorphic or sedimentary rocks requires complete melting. If you do not fully melt the metamorphic or sedimentary rock it will just be a metamorphic rock. The primary way that you can present a rock with high enough temperatures and pressures to melt completely is to bury it to mantle or near mantle depths. This often happens from subduction of an oceanic plate underneath a continental plate. This process formed the Rocky Mountains, Andes, and Aleutian and Japan Island chains. With complete melting you "reset" the clock and the newly formed igneous rock is a brand new mixture of elements and minerals.

Composition of Rocks

A rock can be defined as a solid substance that occurs naturally because of the effects of three basic geological processes: magma solidification; sedimentation of weathered rock debris; and metamorphism.

Most rocks are composed of minerals. *Minerals* are defined by geologists as naturally occurring inorganic solids that have a crystalline structure and a distinct chemical composition. Of course, the minerals found in the Earth's rocks are produced by a variety of different arrangements of chemical elements. A list of the eight most common elements making up the minerals found in the Earth's rocks is described in Table Below.

Table: Common elements found in the Earth's rocks.

Element	Chemical Symbol	Percent Weight in Earth's Crust
Oxygen	O	46.60
Silicon	Si	27.72
Aluminum	Al	8.13
Iron	Fe	5.00
Calcium	Ca	3.63
Sodium	Na	2.83
Potassium	K	2.59
Magnesium	Mg	2.09

Over 2000 minerals have been identified by earth scientists. The *Elements Group* includes over one hundred known minerals. Many of the minerals in this class are composed of only one element. Geologists sometimes subdivide this group into metal and nonmetal categories. *Gold*, *silver*, and *copper* are examples of metals. The elements sulfur and carbon produce the minerals *sulfur*, *diamonds*, and *graphite* which are nonmetallic.

Figure: Silver.   Figure: Copper.   Figure: Graphite.

The *Sulfide Group* are an economically important class of minerals. Many of these minerals consist of metallic elements in chemical combination with the element sulfur. Most ores of important metals such as mercury (*cinnabar* - HgS), iron (*pyrite* - FeS_2), and lead (*galena* - PbS) are extracted from sulfides. Many of the sulfide minerals are recognized by their metallic luster.

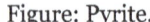
Figure: Pyrite.   Figure: Galena.

The *Halides* are a group of minerals whose principle chemical constituents are fluorine, chlorine, iodine, and bromine. Many of them are very soluble in water. Halides also tend to have a highly ordered molecular structure and a high degree of symmetry. The most well-known mineral of this group is *halite* (NaCl) or rock salt.

Figure: Halite or rock salt.

The *Oxides* are a group of minerals that are compounds of one or more metallic elements combined with oxygen, water, or hydroxyl (OH). The minerals in this mineral group show the greatest variations of physical properties. Some are hard, others soft. Some have a metallic luster, some are clear and transparent. Some representative oxide minerals include *corundum*, *cuprite*, and *hematite*.

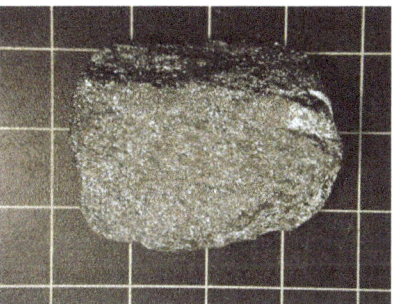

Figure: Corundum   Figure: Hematite

The *Carbonates Group* consists of minerals which contain one or more metallic elements chemically associated with the compound CO_3. Most carbonates are lightly colored and transparent when relatively pure. All carbonates are soft and brittle. Carbonates also effervesce when exposed to warm hydrochloric acid. Most geologists considered the *Nitrates* and *Borates* as subcategories of the carbonates. Some common carbonate minerals include *calcite*, *dolomite*, and *malachite*.

Figure: Calcite.   Figure: Dolomite.

The *Sulfates* are a mineral group that contain one or more metallic element in combination with the sulfate compound SO_4. All sulfates are transparent to translucent and soft. Most are heavy and some are soluble in water. Rarer sulfates exist containing substitutions for the sulfate compound.

For example, in the *chromates* SO_4 is replaced by the compound CrO_4. Two common sulfates are *anhydrite* and *gypsum*.

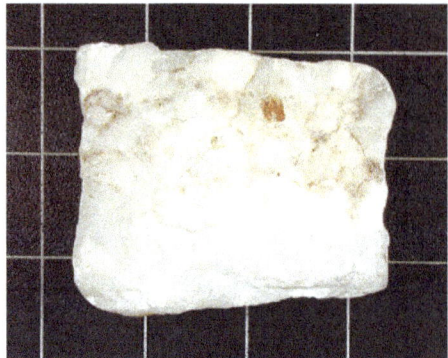

Figure: Gypsum.

The *Phosphates* are a group of minerals of one or more metallic elements chemically associated with the phosphate compound PO_4. The phosphates are often classified together with the arsenate, vanadate, tungstate, and molybdate minerals. One common phosphate mineral is *apatite*. Most phosphates are heavy but soft. They are usually brittle and occur in small crystals or compact aggregates.

The *Silicates* are by far the largest group of minerals. Chemically, these minerals contain varying amounts of silicon and oxygen. It is easy to distinguish silicate minerals from other groups, but difficult to identify individual minerals within this group. None are completely opaque. Most are light in weight. The construction component of all silicates is the tetrahedron. A tetrahedon is a chemical structure where a silicon atom is joined by four oxygen atoms (SiO_4). Some representative minerals include *albite, augite, beryl, biotite, hornblende, microcline, muscovite, olivine, othoclase*, and *quartz*.

Figure: Albite.

Figure: Biotite.

Figure: Hornblende.

Figure: Olivine.

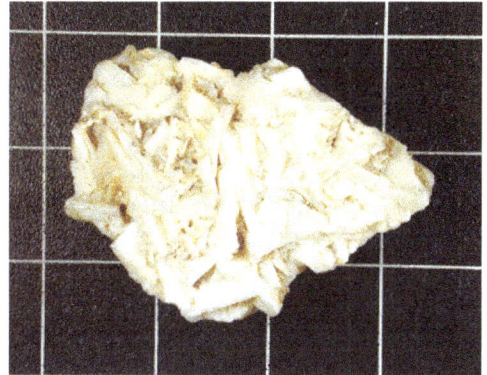

Figure: Orthoclase.

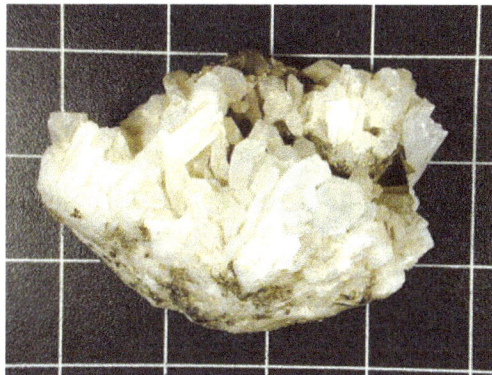

Figure: Quartz.

The *Organic* minerals are a rare group of minerals chemically containing hydrocarbons. Most geologists do not classify these substances as true minerals. Note that our original definition of a mineral excludes organic substances. However, some organic substances that are found naturally on the Earth that exist as crystals that resemble and act like true minerals. These substances are called organic minerals. *Amber* is a good example of an organic mineral.

Igneous Rocks

Igneous rocks form from the cooling of magma – molten materials in the earth's crust. The terminology Igneous means fire or heat. In this sense, igneous rocks are formed when molten rock (magma) solidifies either underneath the earth crust to form plutonic (intrusive) igneous rocks or on the surface of the earth to form volcanic (extrusive) igneous rocks.

They are simply the rocks formed through heating then followed by cooling. The heated material is the molten rock which is made up of partial or complete melting of previously existent rocks in the earth's crust that are consistently subjected to intense heat, high pressure changes, and alterations in composition.

Formation of Igneous Rocks

Molten materials are found below the earth crust and are normally subjected to extreme pressure and temperatures – up to 1200° Celsius. Because of the high temperatures and pressure changes, the molten materials sometimes shoot up to the surface in the form of volcanic eruption and they cool down to form volcanic or extrusive igneous rocks.

Alternatively, some of the molten materials may cool underneath the earth surface very slowly to form plutonic or intrusive igneous rocks. It is because of the extreme heat levels and changes in pressure that igneous rocks do not contain organic matter or fossils. The molten minerals interlock and crystallize as the melt cools and form solid materials.

In the long-run, the melt forms a cool hard rock made up of crystals with no open spaces and don't exhibit any desirable grain alignment. The rocks may be made up entirely of one mineral or various minerals, and their sizes are determined by the cooling process. Rapid cooling results in smaller crystals while slow cooling results in large crystals.

Types of Igneous Rocks

Igneous rocks are of two types, intrusive (plutonic rocks) and extrusive (volcanic rocks).

1. Intrusive Igneous Rocks

    Intrusive igneous rocks are formed when the magma cools off slowly under the earth's crust and hardens into rocks. Gabbro and granite are examples of intrusive igneous rocks. Intrusive rocks are very hard in nature and are often coarse-grained.

2. Extrusive Igneous Rocks

    Extrusive igneous rocks are formed when molten magma spill over to the surface as a result of volcanic eruption. The magma on the surface (lava) cools faster on the surface to form igneous rocks that are fine grained. Examples of such kind of rocks include pumice, basalt, or obsidian.

Sedimentary Rocks

Sedimentary rocks are types of rock that are formed by the deposition and subsequent cementation of mineral or organic particles on the floor of oceans or other bodies of water at the Earth's surface. Sedimentation is the collective name for processes that cause these particles to settle in place. The particles that form a sedimentary rock are called sediment, and may be composed of geological detritus (minerals) or biological detritus (organic matter). Before being deposited, the geological detritus was formed by weathering and erosion from the source area, and then transported to the place of deposition by water, wind, ice, mass movement or glaciers, which are called agents of denudation. Biological detritus was formed by bodies and parts (mainly shells) of dead aquatic organisms, as well as their fecal mass, suspended in water and slowly piling up on the floor of water bodies (marine snow). Sedimentation may also occur as dissolved minerals precipitate from water solution.

The sedimentary rock cover of the continents of the Earth's crust is extensive (73% of the Earth's current land surface), but the total contribution of sedimentary rocks is estimated to be only 8% of the total volume of the crust. Sedimentary rocks are only a thin veneer over a crust consisting mainly of igneous and metamorphic rocks. Sedimentary rocks are deposited in layers as strata, forming a structure called bedding. The study of sedimentary rocks and rock strata provides information about the subsurface that is useful for civil engineering, for example in the construction of roads, houses, tunnels, canals or other structures. Sedimentary rocks are also important sources of natural resources like coal, fossil fuels, drinking water or ores.

The study of the sequence of sedimentary rock strata is the main source for an understanding of the Earth's history, including palaeogeography, paleoclimatology and the history of life. The scientific discipline that studies the properties and origin of sedimentary rocks is called sedimentology. Sedimentology is part of both geology and physical geography and overlaps partly with other disciplines in the Earth sciences, such as pedology, geomorphology, geochemistry and structural geology. Sedimentary rocks have also been found on Mars.

Classification Based on Origin

Sedimentary rocks can be subdivided into four groups based on the processes responsible for their formation: clastic sedimentary rocks, biochemical (biogenic) sedimentary rocks, chemical sedimentary rocks, and a fourth category for "other" sedimentary rocks formed by impacts, volcanism, and other minor processes.

Clastic Sedimentary Rocks

Claystone deposited in Glacial Lake Missoula, Montana, United States. Note the very fine and flat bedding, common for distal lacustrine deposition.

Clastic sedimentary rocks are composed of other rock fragments that were cemented by silicate minerals. Clastic rocks are composed largely of quartz, feldspar, rock (lithic) fragments, clay minerals, and mica; any type of mineral may be present, but they in general represent the minerals that exist locally.

Clastic sedimentary rocks, are subdivided according to the dominant particle size. Most geologists use the Udden-Wentworth grain size scale and divide unconsolidated sediment into three fractions: gravel (>2 mm diameter), sand (1/16 to 2 mm diameter), and mud (clay is <1/256 mm and silt is between 1/16 and 1/256 mm). The classification of clastic sedimentary rocks parallels this scheme; conglomerates and breccias are made mostly of gravel, sandstones are made mostly of sand, and mudrocks are made mostly of the finest material. This tripartite subdivision is mirrored by the broad categories of rudites, arenites, and lutites, respectively, in older literature.

The subdivision of these three broad categories is based on differences in clast shape (conglomerates and breccias), composition (sandstones), grain size or texture (mudrocks).

Conglomerates and Breccias

Conglomerates are dominantly composed of rounded gravel, while breccias are composed of dominantly angular gravel.

Sandstones

Sedimentary rock with sandstone in Malta

Sandstone classification schemes vary widely, but most geologists have adopted the Dott scheme, which uses the relative abundance of quartz, feldspar, and lithic framework grains and the abundance of a muddy matrix between the larger grains.

Composition of Framework Grains

The relative abundance of sand-sized framework grains determines the first word in a sandstone name. Naming depends on the dominance of the three most abundant components quartz, feldspar, or the lithic fragments that originated from other rocks. All other minerals are considered accessories and not used in the naming of the rock, regardless of abundance.

- Quartz sandstones have >90% quartz grains.

- Feldspathic sandstones have <90% quartz grains and more feldspar grains than lithic grains.

- Lithic sandstones have <90% quartz grains and more lithic grains than feldspar grains.

Abundance of Muddy Matrix Material Between Sand Grains

When sand-sized particles are deposited, the space between the grains either remains open or is filled with mud (silt and/or clay sized particle).

- "Clean" sandstones with open pore space (that may later be filled with matrix material) are called arenites.

- Muddy sandstones with abundant (>10%) muddy matrix are called wackes.

Six sandstone names are possible using the descriptors for grain composition (quartz-, feldspathic-, and lithic-) and the amount of matrix (wacke or arenite). For example, a quartz arenite would be composed of mostly (>90%) quartz grains and have little or no clayey matrix between the grains, a lithic wacke would have abundant lithic grains and abundant muddy matrix, etc.

Although the Dott classification scheme is widely used by sedimentologists, common names like greywacke, arkose, and quartz sandstone are still widely used by non-specialists and in popular literature.

Mudrocks

Lower Antelope Canyon was carved out of the surrounding sandstone by both mechanical weathering and chemical weathering. Wind, sand, and water from flash flooding are the primary weathering agents.

Mudrocks are sedimentary rocks composed of at least 50% silt- and clay-sized particles. These relatively fine-grained particles are commonly transported by turbulent flow in water or air, and deposited as the flow calms and the particles settle out of suspension.

Most authors presently use the term "mudrock" to refer to all rocks composed dominantly of mud. Mudrocks can be divided into siltstones, composed dominantly of silt-sized particles; mudstones with subequal mixture of silt- and clay-sized particles; and claystones, composed mostly of clay-sized particles. Most authors use "shale" as a term for a fissile mudrock (regardless of grain size) although some older literature uses the term "shale" as a synonym for mudrock.

Biochemical Sedimentary Rocks

Outcrop of Ordovician oil shale (kukersite), northern Estonia

Biochemical sedimentary rocks are created when organisms use materials dissolved in air or water to build their tissue. Examples include:

- Most types of limestone are formed from the calcareous skeletons of organisms such as corals, mollusks, and foraminifera.

- Coal, formed from plants that have removed carbon from the atmosphere and combined it with other elements to build their tissue.

- Deposits of chert formed from the accumulation of siliceous skeletons of microscopic organisms such as radiolaria and diatoms.

Chemical Sedimentary Rocks

Chemical sedimentary rock forms when mineral constituents in solution become supersaturated and inorganically precipitate. Common chemical sedimentary rocks include oolitic limestone and rocks composed of evaporite minerals, such as halite (rock salt), sylvite, barite and gypsum.

"Other" Sedimentary Rocks

This fourth miscellaneous category includes rocks formed by Pyroclastic flows, impact breccias, volcanic breccias, and other relatively uncommon processes.

Compositional Classification Schemes

Alternatively, sedimentary rocks can be subdivided into compositional groups based on their mineralogy:

- Siliciclastic sedimentary rocks, are dominantly composed of silicate minerals. The sediment that makes up these rocks was transported as bed load, suspended load, or by sediment gravity flows. Siliciclastic sedimentary rocks are subdivided into conglomerates and breccias, sandstone, and mudrocks.

- Carbonate sedimentary rocks are composed of calcite (rhombohedral $CaCO_3$), aragonite (orthorhombic $CaCO_3$), dolomite ($CaMg(CO_3)_2$), and other carbonate minerals based on the CO_3^{2-} ion. Common examples include limestone and dolostone.

- Evaporite sedimentary rocks are composed of minerals formed from the evaporation of water. The most common evaporite minerals are carbonates (calcite and others based on CO_3^{2-}), chlorides (halite and others built on Cl^-), and sulfates (gypsum and others built on SO_4^{2-}). Evaporite rocks commonly include abundant halite (rock salt), gypsum, and anhydrite.

- Organic-rich sedimentary rocks have significant amounts of organic material, generally in excess of 3% total organic carbon. Common examples include coal, oil shale as well as source rocks for oil and natural gas.

- Siliceous sedimentary rocks are almost entirely composed of silica (SiO_2), typically as chert, opal, chalcedony or other microcrystalline forms.

- Iron-rich sedimentary rocks are composed of >15% iron; the most common forms are banded iron formations and ironstones.

- Phosphatic sedimentary rocks are composed of phosphate minerals and contain more than 6.5% phosphorus; examples include deposits of phosphate nodules, bone beds, and phosphatic mudrocks.

Deposition and Transformation

Sediment Transport and Deposition

Cross-bedding and scour in a fine sandstone; the Logan Formation
(Mississippian) of Jackson County, Ohio

Sedimentary rocks are formed when sediment is deposited out of air, ice, wind, gravity, or water flows carrying the particles in suspension. This sediment is often formed when weathering and erosion break down a rock into loose material in a source area. The material is then transported from the source area to the deposition area. The type of sediment transported depends on the geology of the hinterland (the source area of the sediment). However, some sedimentary rocks, such as evaporites, are composed of material that form at the place of deposition. The nature of a sedimentary rock, therefore, not only depends on the sediment supply, but also on the sedimentary depositional environment in which it formed.

Transformation (Diagenesis)

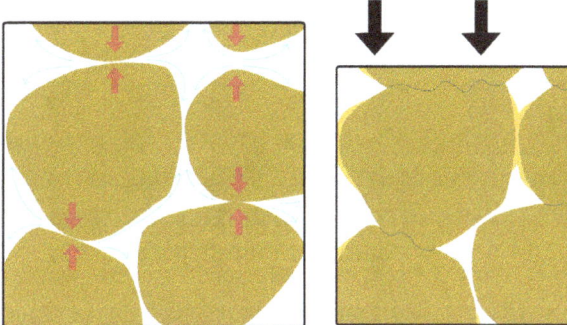

Pressure solution at work in a clastic rock.

While material dissolves at places where grains are in contact, that material may recrystallize from the solution and act as cement in open pore spaces. As a result, there is a net flow of material from areas under high stress to those under low stress, producing a sedimentary rock that is more compact and harder. Loose sand can become sandstone in this way.

The term diagenesis is used to describe all the chemical, physical, and biological changes, exclusive of surface weathering, undergone by a sediment after its initial deposition. Some of those processes cause the sediment to consolidate into a compact, solid substance from the originally loose material. Young sedimentary rocks, especially those of Quaternary age (the most recent period of the geologic time scale) are often still unconsolidated. As sediment deposition builds up, the overburden (lithostatic) pressure rises, and a process known as lithification takes place.

Sedimentary rocks are often saturated with seawater or groundwater, in which minerals can dissolve, or from which minerals can precipitate. Precipitating minerals reduce the pore space in a rock, a process called cementation. Due to the decrease in pore space, the original connate fluids are expelled. The precipitated minerals form a cement and make the rock more compact and competent. In this way, loose clasts in a sedimentary rock can become "glued" together.

When sedimentation continues, an older rock layer becomes buried deeper as a result. The lithostatic pressure in the rock increases due to the weight of the overlying sediment. This causes compaction, a process in which grains mechanically reorganize. Compaction is, for example, an important diagenetic process in clay, which can initially consist of 60% water. During compaction, this interstitial water is pressed out of pore spaces. Compaction can also be the result of dissolution of grains by pressure solution. The dissolved material precipitates again in open pore spaces, which means there is a net flow of material into the pores. However, in some cases, a certain mineral dissolves and does not precipitate again. This process, called leaching, increases pore space in the rock.

Some biochemical processes, like the activity of bacteria, can affect minerals in a rock and are therefore seen as part of diagenesis. Fungi and plants (by their roots) and various other organisms that live beneath the surface can also influence diagenesis.

Burial of rocks due to ongoing sedimentation leads to increased pressure and temperature, which stimulates certain chemical reactions. An example is the reactions by which organic material becomes lignite or coal. When temperature and pressure increase still further, the realm of diagenesis makes way for metamorphism, the process that forms metamorphic rock.

Metamorphic Rocks

Metamorphic rocks are formed through the transformation of pre-existing rocks in a process known as metamorphism (meaning "change in form"). The original rock, or *protolith*, is subjected to heat and pressure which cause physical, chemical and mineralogical changes to the rock. Protoliths may be igneous, sedimentary or pre-existing metamorphic rocks.

Metamorphic rocks are formed within the Earth's crust. Changing temperature and pressure conditions may result in changes to the mineral assemblage of the protolith. Metamorphic rocks are eventually exposed at the surface by uplift and erosion of the overlying rock.

Types of Metamorphic Rocks

Metamorphic rock fall into two categories, foliated and unfoliated. Most foliated metamorphic rocks originate from regional metamorphism. Some unfoliated metamorphic rocks, such as hornfels, originate only by contact metamorphism, but others can originate either by contact metamorphism or by regional metamorphism. Quartz and marble are prime examples of unfoliated that can be produced by either regional or contact metamorphism. Both rock types consist of metamorphic minerals that do not have flat or elongate shapes and thus cannot become layered even if they are produced under differential stress.

A geologist working with metamorphic rocks collects the rocks in the field and looks for the patterns the rocks form in outcrops as well as how those outcrops are related to other types of rock with which they are in contact. Field evidence is often required to know for sure whether rocks are products of regional metamorphism, contact metamorphism, or some other type of metamorphism. If only looking at rock samples in a laboratory, one can be sure of the type of metamorphism that produced a foliated metamorphic rock such as schist or gneiss, or a hornfels, which is unfoliated, but one cannot be sure of the type of metamorphism that produced an unfoliated marble or quartzite.

Foliated Metamorphic Rocks

Foliated metamorphic rocks are named for their style of foliation. However, a more complete name of each particular type of foliated metamorphic rock includes the main minerals that the rock comprises, such as biotite-garnet schist rather than just schist.

- Slate—slates form at low metamorphic grade by the growth of fine-grained chlorite and clay minerals. The preferred orientation of these sheet silicates causes the rock to easily break along parallel planes, giving the rock a slaty cleavage. Some slate breaks into such extensively flat sheets of rock that it is used as the base of pool tables, beneath a layer of rubber and felt. Roof tiles are also sometimes made of slate.

- Phyllite—phyllite is a low-medium grade regional metamorphic rock in which the clay minerals and chlorite have been at least partly replaced by mica mica minerals, muscovite and biotite. This gives the surfaces of phyllite a satiny luster, much brighter than the surface of a piece of slate. It is also common for the differential stresses under which phyllite forms to have produced a set of folds in the rock, making the foliation surfaces wavy or irregular, in contrast to the often perfectly flat surfaces of slaty cleavage.

- Schist—the size of mineral crystals tends to grow larger with increasing metamorphic grade. Schist is a product of medium grades of metamorphism and is characterized by visibly prominent, parallel sheets of mica or similar sheet silicates, usually either muscovite or biotite, or both. In schist, the sheets of mica are usually arranged in irregular planes rather than perfectly flat planes, giving the rock a schistose foliation (or simply schistosity). Schist often contains more than just micas among its minerals, such as quartz, feldspars, and garnet.

- Amphibolite—a poorly foliated to unfoliated mafic metamorphic rock, usually consisting largely of the common black amphibole known as hornblende, plus plagioclase, plus or minus biotite and possibly other minerals; it usually does not contain any quartz. Amphibolite forms at medium-high metamorphic grades.

- Gneiss—like the word schist, the word gneiss is originated from the German language; it is pronounced "nice." As metamorphic grade continue to increase, sheet silicates become unstable and dark minerals such as hornblende or pyroxene start to grow. The dark-colored minerals tend to form separate bands or stripes in the rock, giving it a gneissic foliation of dark and light streaks. Gneiss is a high-grade metamorphic rock. Many types of gneiss look somewhat like granite, except that the gneiss has dark and light stripes whereas in granite randomly oriented and distributed minerals with no stripes or layers.

- Migmatite—a combination of high-grade regional metamorphic rock – usually gneiss or schist – and granitic igneous rock. The granitic rock in migmatite probably originated from partial melting of some of the metamorphic rock, though in some migmatites the granite may have intruded the rock from deeper in the crust. In migmatite you can see metamorphic rock that has reached the limits of metamorphism and begun transitioning into the igneous stage of the rock cycle by melting to form magma.

Names of different styles of foliation come from the common rocks that exhibit such foliation:

- Slate has *slaty* foliation

- Phyllite has *phyllitic* foliation

- Schist has *schistose* foliation

- Gneiss has *gneissic* foliation (also called gneissose foliation)

Nonfoliated Metamorphic Rocks

Nonfoliated metamorphic rocks lack a planar (oriented) fabric, either because the minerals did not grow under differential stress, or because the minerals that grew during metamorphism are not minerals that have elongate or flat shapes. Because they lack foliation, these rocks are named entirely on the basis of their mineralogy.

- Hornfels—hornfels are very hard rocks formed by contact metamorphism of shale, siltstone, or sandstone. The heat from the nearby magma "bakes" the sedimentary rocks and recrystallizes the minerals in them into a new texture that no longer breaks easily along the original sedimentary bedding planes. Depending on the composition of the rock and the temperature reached, minerals indicative of high metamorphic grade such as pyroxene may occur in some hornfels, though many hornfels have minerals indicating medium grade metamorphism.

- Amphibolite—amphibolites are dark-colored rocks with amphibole, usually the common black amphibole known as hornblende, as their most abundant mineral, along with plagioclase and possibly other minerals, though usually no quartz. Amphibolites are poorly foliated to unfoliated and form at medium to medium-high grades of metamorphism from basalt or gabbro.

- Quartzite—quartzite is a metamorphic rock made almost entirely of quartz, for which the protolith was quartz arenite. Because quartz is stable over a wide range of pressure and temperature, little or no new minerals form in quartzite during metamorphism. Instead, the quartz grains recrystallize into a denser, harder rock than the original sandstone. If struck by a rock hammer, quartzite will commonly break right through the quartz grains, rather than around them as when quartz arenite is broken.

- Marble—marble is a metamorphic rock made up almost entirely of either calcite or dolomite, for which the protolith was either limestone or dolostone, respectively. Marbles may have bands of different colors which were deformed into convoluted folds while the rock was ductile. Such marble is often used as decorative stone in buildings. Some marble, which is considered better quality stone for carving into statues, lacks color bands.

Plate Tectonics

Plate tectonics describes the large scale motions of Earth's lithosphere. The theory encompasses the older concepts of continental drift, developed during the first half of the twentieth century, and seafloor spreading, understood during the 1960s.

The outermost part of the Earth's interior is made up of two layers: above is the lithosphere, comprising the crust and the rigid uppermost part of the mantle. Below the lithosphere lies the asthenosphere. Although solid, the asthenosphere has relatively low viscosity and shear strength and can flow like a liquid on geological time scales. The deeper mantle below the asthenosphere is more rigid again due to the higher pressure.

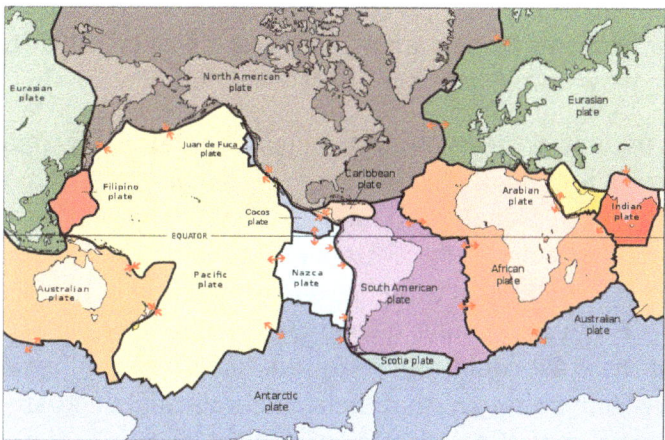

The tectonic plates of the world were mapped in the second half of the twentieth century.

The lithosphere is broken up into what are called *tectonic plates* —in the case of Earth, there are seven major and many minor plates. The lithospheric plates ride on the asthenosphere. These plates move in relation to one another at one of three types of plate boundaries: convergent or collision boundaries, divergent or spreading boundaries, and transform boundaries. Earthquakes, volcanic activity, mountain-building, and oceanic trench formation occur along plate boundaries. The lateral movement of the plates is typically at speeds of 50—100 mm/a.

Synopsis of the Development of the Theory

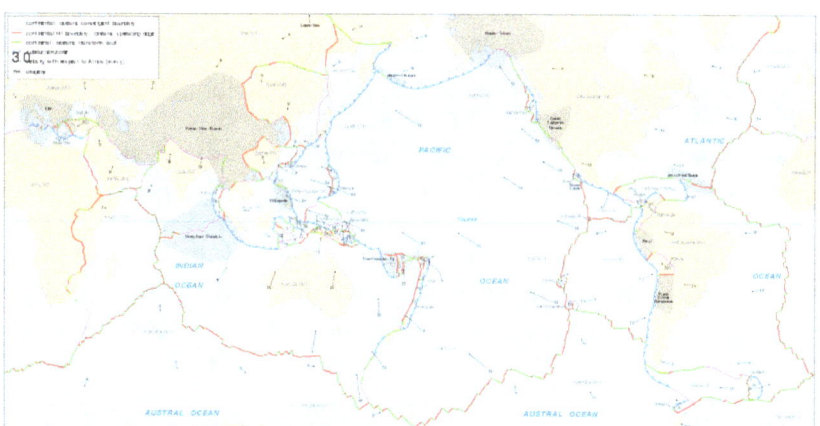

Detailed map showing the tectonic plates with their movement vectors.

In the late nineteenth and early twentieth centuries, geologists assumed that the Earth's major features were fixed, and that most geologic features such as mountain ranges could be explained by vertical crustal movement, as explained by geosynclinal theory. It was observed as early as 1596 that the opposite coasts of the Atlantic Ocean —or, more precisely, the edges of the continental shelves —have similar shapes and seem once to have fitted together. Since that time many theories were proposed to explain this apparent compatibility, but the assumption of a solid earth made the various proposals difficult to explain.

The discovery of radium and its associated heating properties in 1896 prompted a re-examination of the apparent age of the Earth, Those calculations implied that, even if it started at red heat, the

Earth would have dropped to its present temperature in a few tens of millions of years. Armed with the knowledge of a new heat source, scientists reasoned it was credible that the Earth was much older, and also that its core was still sufficiently hot to be liquid.

Plate tectonic theory arose out of the hypothesis of continental drift proposed by Alfred Wegener in 1912 and expanded in his 1915 book *The Origin of Continents and Oceans*. He suggested that the present continents once formed a single land mass that drifted apart, thus releasing the continents from the Earth's core and likening them to "icebergs" of low density granite floating on a sea of more dense basalt. But without detailed evidence and calculation of the forces involved, the theory remained sidelined. The Earth might have a solid crust and a liquid core, but there seemed to be no way that portions of the crust could move around. Later science proved theories proposed by English geologist Arthur Holmes in 1920 that their junctions might actually lie beneath the sea and Holmes' 1928 suggestion of convection currents within the mantle as the driving force.

The first evidence that crust plates did move around came with the discovery of variable magnetic field direction in rocks of differing ages, first revealed at a symposium in Tasmania in 1956. Initially theorized as an expansion of the global crust, later collaborations developed the plate tectonics theory, which accounted for spreading as the consequence of new rock upwelling, but avoided the need for an expanding globe by recognizing subduction zones and conservative translation faults. It was at this point that Wegener's theory moved from radical to mainstream, and became accepted by the scientific community. Additional work on the association of seafloor spreading and magnetic field reversals by Harry Hess and Ron G. Mason pinpointed the precise mechanism which accounted for new rock upwelling.

Following the recognition of magnetic anomalies defined by symmetric, parallel stripes of similar magnetization on the seafloor on either side of a mid-ocean ridge, plate tectonics quickly became broadly accepted. Simultaneous advances in early seismic imaging techniques in and around Wadati-Benioff zones collectively with numerous other geologic observations soon solidified plate tectonics as a theory with extraordinary explanatory and predictive power.

Study of the deep ocean floor was critical to development of the theory; the field of deep sea marine geology accelerated in the 1960s. Correspondingly, plate tectonic theory was developed during the late 1960s and has since been accepted all but universally by scientists throughout all geoscientific disciplines. The theory revolutionized the Earth sciences, explaining a diverse range of geological phenomena and their implications in other studies such as paleogeography and paleobiology.

Key Principles

The division of the outer parts of the Earth's interior into lithosphere and asthenosphere is based on mechanical differences and in the ways that heat is transferred. The lithosphere is cooler and more rigid, while the asthenosphere is hotter and mechanically weaker. Also, the lithosphere loses heat by conduction whereas the asthenosphere also transfers heat by convection and has a nearly adiabatic temperature gradient. This division should not be confused with the *chemical* subdivision of the Earth into (from innermost to outermost) core, mantle, and crust. The lithosphere contains both crust and some mantle. A given piece of mantle may be part of the lithosphere or the asthenosphere at different times, depending on its temperature, pressure and shear strength. The key principle of plate tectonics is that the lithosphere exists as separate and distinct *tectonic plates,* which ride on the fluid-like (visco-elastic solid) asthenosphere. Plate motions range up to a typical 10-40 mm/a (Mid-Atlantic Ridge; about as fast as fingernails grow), to about 160 mm/a (Nazca Plate; about as fast as hair grows).

The plates are around 100 km (60 miles) thick and consist of lithospheric mantle overlain by either of two types of crustal material: oceanic crust (in older texts called *sima* from silicon and magnesium) and continental crust (*sial* from silicon and aluminum). The two types of crust differ in thickness, with continental crust considerably thicker than oceanic (50 km vs. 5 km).

One plate meets another along a *plate boundary,* and plate boundaries are commonly associated with geological events such as earthquakes and the creation of topographic features like mountains, volcanoes and oceanic trenches. The majority of the world's active volcanoes occur along plate boundaries, with the Pacific Plate's Ring of Fire being most active and most widely known.

Tectonic plates can include continental crust or oceanic crust, and a single plate typically carries both. For example, the African Plate includes the continent and parts of the floor of the Atlantic and Indian Oceans. The distinction between continental crust and oceanic crust is based on the density of constituent materials; oceanic crust is denser than continental crust owing to their different proportions of various elements, particularly silicon. Oceanic crust is denser because it has less silicon and more heavier elements ("mafic") than continental crust ("felsic"). As a result, oceanic crust generally lies below sea level (for example most of the Pacific Plate), while the continental crust projects above sea level.

Types of Plate Boundaries

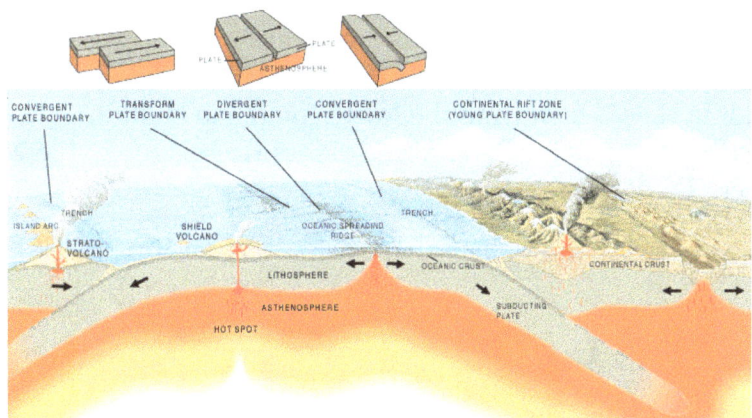

Three types of plate boundary.

Three types of plate boundaries exist, characterized by the way the plates move relative to each other. They are associated with different types of surface phenomena. The different types of plate boundaries are:

1. Transform boundaries occur where plates slide or, perhaps more accurately, grind past each other along transform faults. The relative motion of the two plates is either sinistral (left side toward the observer) or dextral (right side toward the observer). The San Andreas Fault in California is one example.

2. Divergent boundaries occur where two plates slide apart from each other. Mid-ocean ridges (e.g., Mid-Atlantic Ridge) and active zones of rifting (such as Africa's Great Rift Valley) are both examples of divergent boundaries.

3. Convergent boundaries (or *active margins*) occur where two plates slide towards each other commonly forming either a subduction zone (if one plate moves underneath the other) or a continental collision (if the two plates contain continental crust). Deep marine trenches are typically associated with subduction zones. The subducting slab contains many hydrous minerals, which release their water on heating; this water then causes the mantle to melt, producing volcanism. Examples of this are the Andes mountain range in South America and the Japanese island arc.

Transform (Conservative) Boundaries

John Tuzo Wilson recognized that because of friction, the plates cannot simply glide past each other. Rather, stress builds up in both plates and when it reaches a level that exceeds the strain threshold of rocks on either side of the fault the accumulated potential energy is released as strain. Strain is both accumulative and/or instantaneous depending on the rheology of the rock; the ductile lower crust and mantle accumulates deformation gradually via shearing whereas the brittle upper crust reacts by fracture, or instantaneous stress release to cause motion along the fault. The ductile surface of the fault can also release instantaneously when the strain rate is too great. The energy released by instantaneous strain release is the cause of earthquakes, a common phenomenon along transform boundaries.

A good example of this type of plate boundary is the San Andreas Fault which is found in the western coast of North America and is one part of a highly complex system of faults in this area. At this location, the Pacific and North American plates move relative to each other such that the Pacific plate is moving northwest with respect to North America. Other examples of transform faults include the Alpine Fault in New Zealand and the North Anatolian Fault in Turkey. Transform faults are also found offsetting the crests of mid-ocean ridges (for example, the Mendocino Fracture Zone offshore northern California).

Divergent (Constructive) Boundaries

Bridge across the Álfagjá rift valley in southwest Iceland, the boundary between the Eurasian and North American continental tectonic plates.

At divergent boundaries, two plates move apart from each other and the space that this creates is filled with new crustal material sourced from molten magma that forms below. The origin of new divergent boundaries at triple junctions is sometimes thought to be associated with the phenomenon known as hotspots. Here, exceedingly large convective cells bring very large quantities of hot asthenospheric material near the surface and the kinetic energy is thought to be sufficient to break apart the lithosphere. The hot spot which may have initiated the Mid-Atlantic Ridge system currently underlies Iceland which is widening at a rate of a few centimeters per year.

Divergent boundaries are typified in the oceanic lithosphere by the rifts of the oceanic ridge system, including the Mid-Atlantic Ridge and the East Pacific Rise, and in the continental lithosphere by rift valleys such as the famous East African Great Rift Valley. Divergent boundaries can create massive fault zones in the oceanic ridge system. Spreading is generally not uniform, so where spreading rates of adjacent ridge blocks are different, massive transform faults occur. These are the fracture zones, many bearing names, that are a major source of submarine earthquakes. A sea floor map will show a rather strange pattern of blocky structures that are separated by linear features perpendicular to the ridge axis. If one views the sea floor between the fracture zones as conveyor belts carrying the ridge on each side of the rift away from the spreading center the action becomes clear. Crest depths of the old ridges, parallel to the current spreading center, will be older and deeper (from thermal contraction and subsidence).

It is at mid-ocean ridges that one of the key pieces of evidence forcing acceptance of the sea-floor spreading hypothesis was found. Airborne geomagnetic surveys showed a strange pattern of symmetrical magnetic reversals on opposite sides of ridge centers. The pattern was far too regular to be coincidental as the widths of the opposing bands were too closely matched. Scientists had been studying polar reversals and the link was made by Lawrence W. Morley, Frederick John Vine and Drummond Hoyle Matthews in the Morley-Vine-Matthews hypothesis. The magnetic banding directly corresponds with the Earth's polar reversals. This was confirmed by measuring the ages of the rocks within each band. The banding furnishes a map in time and space of both spreading rate and polar reversals.

Convergent (Destructive) Boundaries

The nature of a convergent boundary depends on the type of lithosphere in the plates that are colliding. Where a dense oceanic plate collides with a less-dense continental plate, the oceanic plate is typically thrust underneath because of the greater buoyancy of the continental lithosphere, forming a subduction zone. At the surface, the topographic expression is commonly an oceanic trench on the ocean side and a mountain range on the continental side. An example of a continental-oceanic subduction zone is the area along the western coast of South America where the oceanic Nazca Plate is being subducted beneath the continental South American Plate.

While the processes directly associated with the production of melts directly above downgoing plates producing surface volcanism is the subject of some debate in the geologic community, the general consensus from ongoing research suggests that the release of volatiles is the primary contributor. As the subducting plate descends, its temperature rises driving off volatiles (most importantly water) encased in the porous oceanic crust. As this water rises into the mantle of the overriding plate, it lowers the melting temperature of surrounding mantle, producing melts (magma) with large amounts of dissolved gases. These melts rise to the surface and are the source of some of the most explosive volcanism on Earth because of their high volumes of extremely pressurized gases (consider Mount St. Helens). The melts rise to the surface and cool forming long chains of volcanoes inland from the continental shelf and parallel to it. The continental spine of western South America is dense with this type of volcanic mountain building from the subduction of the Nazca plate. In North America the Cascade mountain range, extending north from California's Sierra Nevada, is also of this type. Such volcanoes are characterized by alternating periods of quiet and episodic eruptions that start with explosive gas expulsion with fine particles of glassy volcanic ash and spongy cinders, followed by a rebuilding phase with hot magma. The entire Pacific Ocean boundary is surrounded by long stretches of volcanoes and is known collectively as *The Ring of Fire.*

Where two continental plates collide the plates either buckle and compress or one plate delves under or (in some cases) overrides the other. Either action will create extensive mountain ranges. The most dramatic effect seen is where the northern margin of the Indian Plate is being thrust under a portion of the Eurasian plate, lifting it and creating the Himalayas and the Tibetan Plateau beyond. It may have also pushed nearby parts of the Asian continent aside to the east.

When two plates with oceanic crust converge they typically create an island arc as one plate is subducted below the other. The arc is formed from volcanoes which erupt through the overriding plate as the descending plate melts below it. The arc shape occurs because of the spherical surface of the earth (nick the peel of an orange with a knife and note the arc formed by the straight-edge of the knife). A deep undersea trench is located in front of such arcs where the descending slab dips downward. Good examples of this type of plate convergence would be Japan and the Aleutian Islands in Alaska.

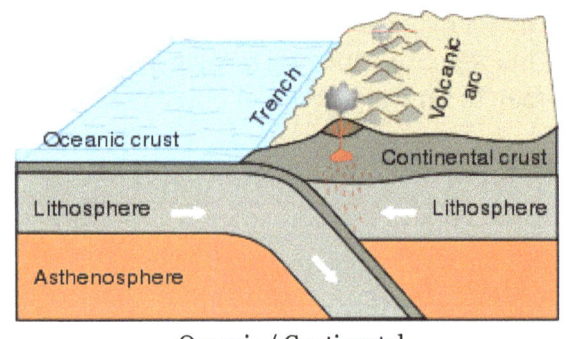

Oceanic / Continental

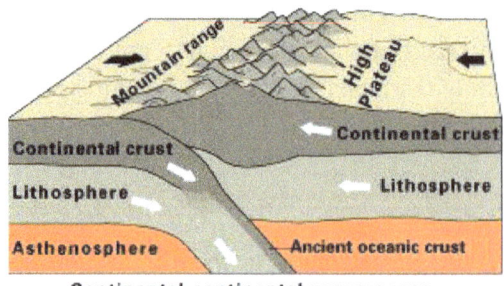

Continental / Continental

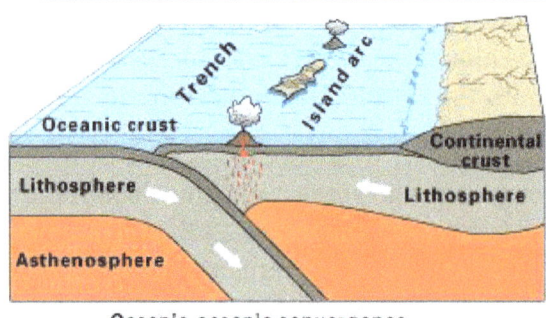

Oceanic / Oceanic

Plates may collide at an oblique angle rather than head-on to each other (e.g. one plate moving north, the other moving south-east), and this may cause strike-slip faulting along the collision zone, in addition to subduction or compression.

Not all plate boundaries are easily defined. Some are broad belts whose movements are unclear to scientists. One example would be the Mediterranean-Alpine boundary, which involves two major plates and several micro plates. The boundaries of the plates do not necessarily coincide with those of the continents. For instance, the North American Plate covers not only North America, but also far northeastern Siberia, plus a substantial portion of the Atlantic Ocean.

Driving Forces of Plate Motion

Tectonic plates are able to move because of the relative density of oceanic lithosphere and the relative weakness of the asthenosphere. Dissipation of heat from the mantle is acknowledged to be the original source of energy driving plate tectonics. The current view, although it is still a matter of some debate, is that excess density of the oceanic lithosphere sinking in subduction zones is the most powerful source of plate motion. When it forms at mid-ocean ridges, the oceanic lithosphere is initially less dense than the underlying asthenosphere, but it becomes more dense with age, as it conductively cools and thickens. The greater density of old lithosphere relative to the underlying asthenosphere allows it to sink into the deep mantle at subduction zones, providing most of the driving force for plate motions. The weakness of the asthenosphere allows the tectonic plates to move easily towards a subduction zone. Although subduction is believed to be the strongest force driving plate motions, it cannot be the only force since there are plates such as the North American Plate which are moving, yet are nowhere being subducted. The same is true for the enormous Eurasian Plate. The sources of plate motion are a matter of intensive research and discussion among earth scientists.

Two and three-dimensional imaging of the Earth's interior (seismic tomography) shows that there is a laterally heterogeneous density distribution throughout the mantle. Such density variations can be material (from rock chemistry), mineral (from variations in mineral structures), or thermal (through thermal expansion and contraction from heat energy). The manifestation of this lateral density heterogeneity is mantle convection from buoyancy forces. How mantle convection relates directly and indirectly to the motion of the plates is a matter of ongoing study and discussion in geodynamics. Somehow, this energy must be transferred to the lithosphere in order for tectonic plates to move. There are essentially two types of forces that are thought to influence plate motion: friction and gravity.

Friction

Basal Drag

Large scale convection currents in the upper mantle are transmitted through the asthenosphere; motion is driven by friction between the asthenosphere and the lithosphere.

Slab Suction

Local convection currents exert a downward frictional pull on plates in subduction zones at ocean trenches. Slab suction may occur in a geodynamic setting wherein basal tractions continue to act on the plate as it dives into the mantle (although perhaps to a greater extent acting on both the under and upper side of the slab).

Gravitation

Gravitational sliding: Plate motion is driven by the higher elevation of plates at ocean ridges. As oceanic lithosphere is formed at spreading ridges from hot mantle material it gradually cools and thickens with age (and thus distance from the ridge). Cool oceanic lithosphere is significantly denser than the hot mantle material from which it is derived and so with increasing thickness it gradually subsides into the mantle to compensate the greater load. The result is a slight lateral incline with distance from the ridge axis.

Casually in the geophysical community and more typically in the geological literature in lower education this process is often referred to as "ridge-push." This is, in fact, a misnomer as nothing is "pushing" and tensional features are dominant along ridges. It is more accurate to refer to this mechanism as gravitational sliding as variable topography across the totality of the plate can vary considerably and the topography of spreading ridges is only the most prominent feature. For example:

1. Flexural bulging of the lithosphere before it dives underneath an adjacent plate, for instance, produces a clear topographical feature that can offset or at least affect the influence of topographical ocean ridges.

2. Mantle plumes impinging on the underside of tectonic plates can drastically alter the topography of the ocean floor.

Slab-pull

Plate motion is partly driven by the weight of cold, dense plates sinking into the mantle at trenches. There is considerable evidence that convection is occurring in the mantle at some scale. The upwelling of material at mid-ocean ridges is almost certainly part of this convection. Some early models of plate tectonics envisioned the plates riding on top of convection cells like conveyor belts. However, most scientists working today believe that the asthenosphere is not strong enough to directly cause motion by the friction of such basal forces. Slab pull is most widely thought to be the greatest force acting on the plates. Recent models indicate that trench suction plays an important role as well. However, it should be noted that the North American Plate, for instance, is nowhere being subducted, yet it is in motion. Likewise the African, Eurasian and Antarctic Plates. The overall driving force for plate motion and its energy source remain subjects of ongoing research.

External Forces

In a study published in the January-February 2006 issue of the *Geological Society of America Bulletin,* a team of Italian and U.S. scientists argued that the westward component of plates is from Earth's rotation and consequent tidal friction of the Moon. As the Earth spins eastward beneath the moon, they say, the moon's gravity ever so slightly pulls the Earth's surface layer back westward. It has also been suggested (albeit, controversially) that this observation may also explain why Venus and Mars have no plate tectonics since Venus has no moon, and Mars' moons are too small to have significant tidal effects on Mars. This is not, however, a new argument.

It was originally raised by the "father" of the plate tectonics hypothesis, Alfred Wegener. It was challenged by the physicist Harold Jeffreys who calculated that the magnitude of tidal friction required would have quickly brought the Earth's rotation to a halt long ago. Many plates are moving north and eastward, and the dominantly westward motion of the Pacific ocean basins is simply from the eastward bias of the Pacific spreading center (which is not a predicted manifestation of such lunar forces). It is argued, however, that relative to the lower mantle, there is a slight westward component in the motions of all the plates.

Relative Significance of each Mechanism

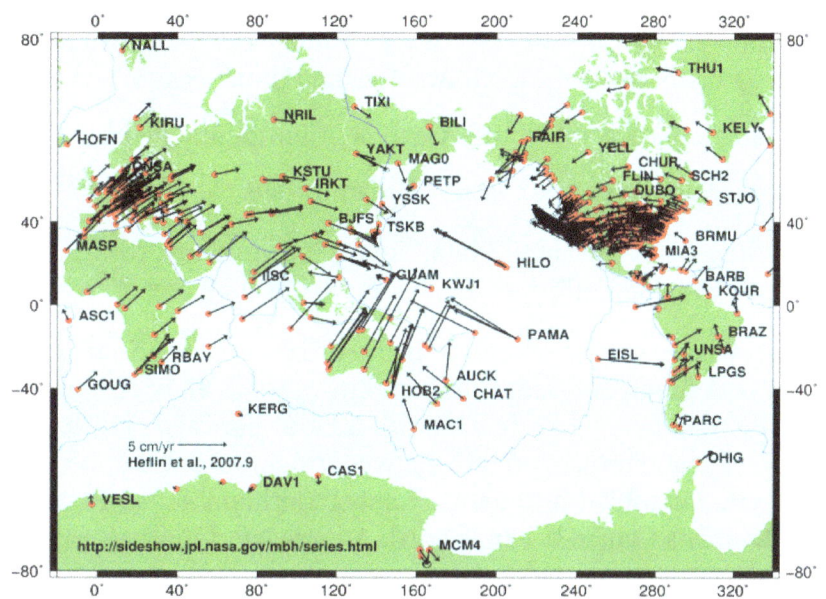

Plate motion based on Global Positioning System (GPS) satellite data from NASA JPL.
Vectors show direction and magnitude of motion.

The actual vector of a plate's motion must necessarily be a function of all the forces acting upon the plate. However, therein remains the problem regarding what degree each process contributes to the motion of each tectonic plate.

The diversity of geodynamic settings and properties of each plate must clearly result in differences in the degree to which such processes are actively driving the plates. One method of dealing with this problem is to consider the relative rate at which each plate is moving and to consider the available evidence of each driving force upon the plate as far as possible.

One of the most significant correlations found is that lithospheric plates attached to downgoing (subducting) plates move much faster than plates not attached to subducting plates. The Pacific plate, for instance, is essentially surrounded by zones of subduction (the so-called Ring of Fire) and moves much faster than the plates of the Atlantic basin, which are attached (perhaps one could say 'welded') to adjacent continents instead of subducting plates. It is thus thought that forces associated with the downgoing plate (slab pull and slab suction) are the driving forces which determine the motion of plates, except for those plates which are not being subducted.

The driving forces of plate motion are, nevertheless, still very active subjects of on-going discussion and research in the geophysical community.

Major Plates

The main plates are:

- African Plate covering Africa - Continental plate

- Antarctic Plate covering Antarctica - Continental plate

- Australian Plate covering Australia - Continental plate

- Indian Plate covering Indian subcontinent and a part of Indian Ocean - Continental plate

- Eurasian Plate covering Asia and Europe - Continental plate

- North American Plate covering North America and north-east Siberia - Continental plate

- South American Plate covering South America - Continental plate

- Pacific Plate covering the Pacific Ocean - Oceanic plate

Notable minor plates include the Arabian Plate, the Caribbean Plate, the Juan de Fuca Plate, the Cocos Plate, the Nazca Plate, the Philippine Plate and the Scotia Plate.

The movement of plates has caused the formation and break-up of continents over time, including occasional formation of a supercontinent that contains most or all of the continents. The supercontinent Rodinia is thought to have formed about 1 billion years ago and to have embodied most or all of Earth's continents, and broken up into eight continents around 600 million years ago. The eight continents later re-assembled into another supercontinent called Pangaea; Pangaea eventually broke up into Laurasia (which became North America and Eurasia) and Gondwana (which became the remaining continents).

Historical Development of the Theory

Continental Drift

Continental drift was one of many ideas about tectonics proposed in the late nineteenth and early twentieth centuries. The theory has been superseded and the concepts and data have been incorporated within plate tectonics.

By 1915, Alfred Wegener was making serious arguments for the idea in the first edition of *The Origin of Continents and Oceans*. In that book, he noted how the east coast of South America and the west coast of Africa looked as if they were once attached. Wegener wasn't the first to note this (Abraham Ortelius, Francis Bacon, Benjamin Franklin, Snider-Pellegrini, Roberto Mantovani and Frank Bursley Taylor preceded him), but he was the first to marshal significant fossil and paleo-topographical and climatological evidence to support this simple observation (and was supported in this by researchers such as Alex du Toit). However, his ideas were not taken seriously by many geologists, who pointed out that there was no apparent mechanism for continental drift. Specifically, they did not see how continental rock could plow through the much denser rock that makes up oceanic crust. Wegener could not explain the force that propelled continental drift.

Wegener's vindication did not come until after his death in 1930. In 1947, a team of scientists led by Maurice Ewing utilizing the Woods Hole Oceanographic Institution's research vessel *Atlantis* and an array of instruments, confirmed the existence of a rise in the central Atlantic Ocean, and found that the floor of the seabed beneath the layer of sediments consisted of basalt, not the granite which is the main constituent of continents. They also found that the oceanic crust was much thinner than continental crust. All these new findings raised important and intriguing questions.

Beginning in the 1950s, scientists including Harry Hess, using magnetic instruments (magnetometers) adapted from airborne devices developed during World War II to detect submarines, began recognizing odd magnetic variations across the ocean floor. This finding, though unexpected, was not entirely surprising because it was known that basalt—the iron-rich, volcanic rock making up the ocean floor—contains a strongly magnetic mineral (magnetite) and can locally distort compass readings. This distortion was recognized by Icelandic mariners as early as the late eighteenth century. More important, because the presence of magnetite gives the basalt measurable magnetic properties, these newly discovered magnetic variations provided another means to study the deep ocean floor. When newly formed rock cools, such magnetic materials recorded the Earth's magnetic field at the time.

As more and more of the seafloor was mapped during the 1950s, the magnetic variations turned out not to be random or isolated occurrences, but instead revealed recognizable patterns. When these magnetic patterns were mapped over a wide region, the ocean floor showed a zebra-like pattern. Alternating stripes of magnetically different rock were laid out in rows on either side of the mid-ocean ridge: one stripe with normal polarity and the adjoining stripe with reversed polarity. The overall pattern, defined by these alternating bands of normally and reversely polarized rock, became known as magnetic striping.

When the rock strata of the tips of separate continents are very similar it suggests that these rocks were formed in the same way implying that they were joined initially. For instance, some parts of Scotland and Ireland contain rocks very similar to those found in Newfoundland and New Brunswick. Furthermore, the Caledonian Mountains of Europe and parts of the Appalachian Mountains of North America are very similar in structure and lithology.

Floating Continents

The prevailing concept was that there were static shells of strata under the continents. It was observed early that although granite existed on continents, seafloor seemed to be composed of denser basalt. It was apparent that a layer of basalt underlies continental rocks.

However, based upon abnormalities in plumb line deflection by the Andes in Peru, Pierre Bouguer deduced that less-dense mountains must have a downward projection into the denser layer underneath. The concept that mountains had "roots" was confirmed by George B. Airy a hundred years later during study of Himalayan gravitation, and seismic studies detected corresponding density variations.

By the mid-1950s the question remained unresolved of whether mountain roots were clenched in surrounding basalt or were floating like an iceberg.

In 1958 the Tasmanian geologist Samuel Warren Carey published an essay *The tectonic approach to continental drift* in support of the expanding earth model.

Plate Tectonic Theory

Significant progress was made in the 1960s, and was prompted by a number of discoveries, most notably the Mid-Atlantic ridge. The most notable was the 1962 publication of a paper by American geologist Harry Hammond Hess. Hess suggested that instead of continents moving *through* oceanic crust (as was suggested by continental drift) that an ocean basin and its adjoining continent moved together on the same crustal unit, or plate. In the same year, Robert R. Coats of the U.S. Geological Survey described the main features of island arc subduction in the Aleutian Islands. His paper, though little-noted (and even ridiculed) at the time, has since been called "seminal" and "prescient." In 1967, W. Jason Morgan proposed that the Earth's surface consists of 12 rigid plates that move relative to each other. Two months later, in 1968, Xavier Le Pichon published a complete model based on 6 major plates with their relative motions.

Explanation of Magnetic Striping

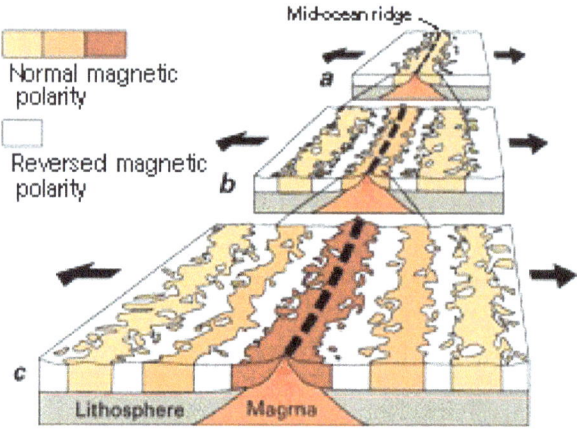

Seafloor magnetic striping.

The discovery of magnetic striping and the stripes being symmetrical around the crests of the mid-ocean ridges suggested a relationship. In 1961, scientists began to theorize that mid-ocean ridges mark structurally weak zones where the ocean floor was being ripped in two lengthwise along the ridge crest. New magma from deep within the Earth rises easily through these weak zones and eventually erupts along the crest of the ridges to create new oceanic crust. This process, later called seafloor spreading, operating over many millions of years continues to form new ocean floor all across the 50,000 km-long system of mid-ocean ridges. This hypothesis was supported by several lines of evidence:

1. at or near the crest of the ridge, the rocks are very young, and they become progressively older away from the ridge crest;

2. the youngest rocks at the ridge crest always have present-day (normal) polarity;

3. stripes of rock parallel to the ridge crest alternated in magnetic polarity (normal-reversed-normal, etc.), suggesting that the Earth's magnetic field has reversed many times.

By explaining both the zebra like magnetic striping and the construction of the mid-ocean ridge system, the seafloor spreading hypothesis quickly gained converts and represented another major advance in the development of the plate-tectonics theory. Furthermore, the oceanic crust now came to be appreciated as a natural "tape recording" of the history of the reversals in the Earth's magnetic field.

Subduction Discovered

A profound consequence of seafloor spreading is that new crust was, and is now, being continually created along the oceanic ridges. This idea found great favor with some scientists, most notably S. Warren Carey, who claimed that the shifting of the continents can be simply explained by a large increase in size of the Earth since its formation. However, this so-called "Expanding Earth theory" hypothesis was unsatisfactory because its supporters could offer no convincing mechanism to produce a significant expansion of the Earth. Certainly there is no evidence that the moon has expanded in the past 3 billion years. Still, the question remained: how can new crust be continuously added along the oceanic ridges without increasing the size of the Earth?

This question particularly intrigued Harry Hess, a Princeton University geologist and a Naval Reserve Rear Admiral, and Robert S. Dietz, a scientist with the U.S. Coast and Geodetic Survey who first coined the term *seafloor spreading*. Dietz and Hess were among the small handful who really understood the broad implications of sea floor spreading. If the Earth's crust was expanding along the oceanic ridges, Hess reasoned, it must be shrinking elsewhere. He suggested that new oceanic crust continuously spreads away from the ridges in a conveyor belt-like motion. Many millions of years later, the oceanic crust eventually descends into the oceanic trenches —very deep, narrow canyons along the rim of the Pacific Ocean basin. According to Hess, the Atlantic Ocean was expanding while the Pacific Ocean was shrinking. As old oceanic crust is consumed in the trenches, new magma rises and erupts along the spreading ridges to form new crust. In effect, the ocean basins are perpetually being "recycled," with the creation of new crust and the destruction of old oceanic lithosphere occurring simultaneously. Thus, Hess' ideas neatly explained why the Earth does not get bigger with sea floor spreading, why there is so little sediment accumulation on the ocean floor, and why oceanic rocks are much younger than continental rocks.

Mapping with Earthquakes

During the twentieth century, improvements in and greater use of seismic instruments such as seismographs enabled scientists to learn that earthquakes tend to be concentrated in certain areas, most notably along the oceanic trenches and spreading ridges. By the late 1920s, seismologists were beginning to identify several prominent earthquake zones parallel to the trenches that typically were inclined 40–60° from the horizontal and extended several hundred kilometers into the Earth. These zones later became known as Wadati-Benioff zones, or simply Benioff zones, in honor of the seismologists who first recognized them, Kiyoo Wadati of Japan and Hugo Benioff of the United States. The study of global seismicity greatly advanced in the 1960s with the establishment of the Worldwide Standardized Seismograph Network (WWSSN) to monitor the compliance of the 1963 treaty banning above-ground testing of nuclear weapons. The much-improved data from the WWSSN instruments allowed seismologists to map precisely the zones of earthquake concentration world wide.

Geological Paradigm Shift

The acceptance of the theories of continental drift and sea floor spreading (the two key elements of plate tectonics) may be compared to the Copernican revolution in astronomy. Within a matter of only several years geophysics and geology in particular were revolutionized. The parallel is striking: just as pre-Copernican astronomy was highly descriptive but still unable to provide explanations for the motions of celestial objects, pre-tectonic plate geological theories described what was observed but struggled to provide any fundamental mechanisms. The problem lay in the question "How?." Before acceptance of plate tectonics, geology in particular was trapped in a "pre-Copernican" box.

However, by comparison to astronomy the geological revolution was much more sudden. What had been rejected for decades by any respectable scientific journal was eagerly accepted within a few short years in the 1960s and 1970s. Any geological description before this had been highly descriptive. All the rocks were described and assorted reasons, sometimes in excruciating detail, were given for why they were where they are. The descriptions are still valid. The reasons, however, today sound much like pre-Copernican astronomy.

One simply has to read the pre-plate descriptions of why the Alps or Himalaya exist to see the difference. In an attempt to answer "how" questions like "How can rocks that are clearly marine in origin exist thousands of meters above sea-level in the Dolomites?," or "How did the convex and concave margins of the Alpine chain form?," any true insight was hidden by complexity that boiled down to technical jargon without much fundamental insight as to the underlying mechanics.

With plate tectonics answers quickly fell into place or a path to the answer became clear. Collisions of converging plates had the force to lift the sea floor to great heights. The cause of marine trenches oddly placed just off island arcs or continents and their associated volcanoes became clear when the processes of subduction at converging plates were understood.

Mysteries were no longer mysteries. Forests of complex and obtuse answers were swept away. Why were there striking parallels in the geology of parts of Africa and South America? Why did Africa and South America look strangely like two pieces that should fit to anyone having done a jigsaw puzzle? Look at some pre-tectonics explanations for complexity. For simplicity and one that explained a great deal more look at plate tectonics. A great rift, similar to the Great Rift Valley in northeastern Africa, had split apart a single continent, eventually forming the Atlantic Ocean, and the forces were still at work in the Mid-Atlantic Ridge.

We have inherited some of the old terminology, but the underlying concept is as radical and simple as was "The Earth moves" in astronomy.

Biogeographic Implications on Biota

Continental drift theory helps biogeographers to explain the disjunct biogeographic distribution of present day life found on different continents but having similar ancestors. In particular, it explains the Gondwanan distribution of ratites and the Antarctic flora.

Plate Tectonics on other Planets

The appearance of plate tectonics on terrestrial planets is related to planetary mass, with more massive planets than Earth expected to exhibit plate tectonics. Earth may be a borderline case, owing its tectonic activity to abundant water.

Venus

Venus shows no evidence of active plate tectonics. There is debatable evidence of active tectonics in the planet's distant past; however, events taking place since then (such as the plausible and generally accepted hypothesis that the Venusian lithosphere has thickened greatly over the course of several hundred million years) has made constraining the course of its geologic record difficult. However, the numerous well-preserved impact craters have been utilized as a dating method to approximately date the Venusian surface (since there are thus far no known samples of Venusian rock to be dated by more reliable methods). Dates derived are the dominantly in the range ~500 to 750 Ma, although ages of up to ~1.2 Ga have been calculated. This research has led to the fairly well accepted hypothesis that Venus has undergone an essentially complete volcanic resurfacing at least once in its distant past, with the last event taking place approximately within the range of estimated surface ages. While the mechanism of such an impressionable thermal event remains a debated issue in Venusian geosciences, some scientists are advocates of processes involving plate motion to some extent.

One explanation for Venus' lack of plate tectonics is that on Venus temperatures are too high for significant water to be present. The Earth's crust is soaked with water, and water plays an important role in the development of shear zones. Plate tectonics requires weak surfaces in the crust along which crustal slices can move, and it may well be that such weakening never took place on Venus because of the absence of water. However, some researchers remain convinced that plate tectonics is or was once active on this planet.

Mars

Unlike Venus, the crust of Mars has water in it and on it (mostly in the form of ice). This planet is considerably smaller than the Earth, but shows some indications that could suggest a similar style of tectonics. The gigantic volcanoes in the Tharsis area are linearly aligned like volcanic arcs on Earth; the enormous canyon Valles Marineris could have been formed by some form of crustal spreading.

As a result of observations made of the magnetic field of Mars by the *Mars Global Surveyor* spacecraft in 1999, large scale patterns of magnetic striping were discovered on this planet. To explain these magnetization patterns in the Martian crust it has been proposed that a mechanism similar to plate tectonics may once have been active on the planet. Further data from the *Mars Express* orbiter's *High Resolution Stereo Camera* in 2007 clearly showed an example in the Aeolis Mensae region.

Galilean Satellites

Some of the satellites of Jupiter have features that may be related to plate-tectonic style deformation, although the materials and specific mechanisms may be different from plate-tectonic activity on Earth.

Titan

Titan, the largest moon of Saturn, was reported to show tectonic activity in images taken by the Huygens Probe, which landed on Titan on January 14, 2005.

Earthquakes

The sudden shaking or rolling of the earth's surface is called an earthquake. Actually earthquakes occur daily around the world (according to one estimate, about 8000 occur every year), but most of them are too mild to be noticeable. We know of them only because they are recorded by instruments called seismographs (the Greek word seismos means 'earthquake').

Occurrence of Earthquakes:

Perhaps you remember that the earth is made up of three layers. At its heart is a core of iron, consisting of a solid sphere surrounded by a layer of hot, molten iron. Around the core is a mantle of soft, paste like rocks. And over the mantle rests the hard layer of rocks we call the crust. This crust is not a uniform, faultless shell. It is more like a jigsaw of blocks that fit together. The huge blocks that make up the crust are called tectonic plates.

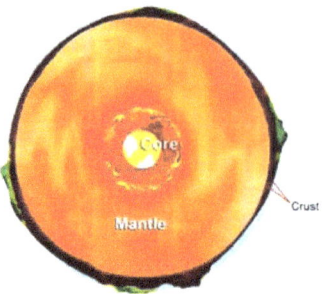

Figure: The plates of the crust float on the soft mantle

The heat inside the earth sets up a current in the mantle, keeping it in constant motion. This makes the plates of the crust move continually, like rafts on a gentle ocean. The movement sometimes causes the edges of the plates to grind against each other with a lot of force.

They may then get deformed, displaced, crushed or fractured. They may also slide under each other or move apart. Such changes in the plates send a tremor or set up vibrations through the crust, causing what we call an earthquake.

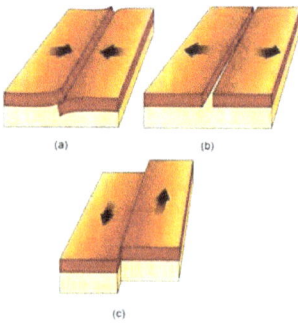

Figure: The edges of tectonic plates may (a) slip under
each other, (b) move apart or (c) get displaced

Earthquakes can occur due to reasons other than plate movements. Volcanic activity can cause earthquakes, as can human activities like nuclear explosions carried out underground. The collapse of mines has also been known to cause minor earthquakes.

Dams:

The build-up of pressure due to the storage of a large amount of water in the reservoirs behind large dams is considered to be a potential cause of earthquakes. The earthquake that occurred in Koyna (Maharashtra) in 1967, for example, is thought to have been caused by the Koyna dam.

The dam was completed in 1963 and several tremors were felt as the reservoir was being filled. The earthquake that occurred in 1967 was quite strong. It killed 200 people and injured 1500. It also caused cracks in the dam.

Measuring an Earthquake:

Earthquakes usually start at a depth of less than 100 km below the ground. The point of origin, called the seismic focus or hypocenter, is located with the help of seismographs. (Seismographs across the world are constantly noting the vibrations of the crust.)

Vibrations spread out from the hypocentre, like ripples in a pool of water. The location on the surface of the earth directly above the hypocenter is called the epicenter. It normally bears the brunt of the destructive power of these vibrations. That is to say, this is where the maximum damage normally occurs.

The extent of the damage depends on the strength of the vibrations or the energy associated with them. It also depends on the density of population (how many people live in an area) and the way buildings are constructed.

The nature of the soil is another factor which determines the extent of damage. If the soil is loose and damp, the damage is greater than if it is hard and firm. This is why the severity of an earthquake is measured in two ways—in terms of its magnitude and in terms of its intensity.

Richter Scale:

The magnitude of an earthquake depends on the energy of the vibrations. It is measured by seismographs on a scale called the Richter scale. The range of this scale is from 0 to 10. The energy of the vibrations increases by steps of about 30 on this scale.

In other words, the vibrations of an earthquake measuring 6 on this scale would be 30 times more energetic than those of a quake measuring 5. Earthquakes measuring 9 or more on this scale are rare. Those measuring from 8 to 8.9 are quite devastating, while those between 7 and 7.9 are considered major. Even moderate (5 to 5.9) and strong (6.0 to 6.9) earthquakes are quite destructive in densely populated areas.

Volcanoes

A volcano is a vent or chimney, which transfers molten rock (known as magma) from depth to the Earth's surface through eruptions. Magma erupting from a volcano is called lava. Lava builds up around the vent and forms a cone.

A volcano is currently active if it is erupting lava, releasing gas or generating seismic activity. An active volcano is labeled dormant if it has not erupted for a long time but could erupt again in the future. When a volcano has been dormant for more than 10 000 years, it is considered extinct. Volcanoes can remain inactive, or dormant, for hundreds or thousands of years before erupting again. During this time, they can become covered by vegetation, making them difficult to identify.

How explosive a volcanic eruption is depends on how easily magma can flow or trap gas. If magma is able to trap a large amount of gas, it can produce explosive eruptions.

Volcanoes can have many different appearances. Some volcanoes are perfect cone shapes while others are deep depressions filled with water. The shape of a volcano provides clues to the type and size of eruption that occurred. Eruption types and sizes depend on what the magma is made up of.

Causes of Volcanoes

Volcanoes occur when material significantly warmer than its surroundings is erupted onto the surface of a planet or moon from its interior. On Earth, the erupted material can be liquid rock ("lava" when it's on the surface, "magma" when it's underground), ash, cinders, and/or gas. There are three reasons why magma might rise and cause eruptions onto Earth's surface.

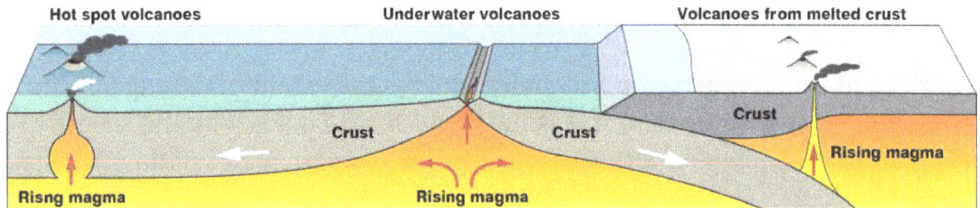

Volcanoes on Earth form from rising magma. Magma rises in three different ways.

Magma can rise when pieces of Earth's crust called tectonic plates slowly move away from each other. The magma rises up to fill in the space. When this happens underwater volcanoes can form.

Magma also rises when these tectonic plates move toward each other. When this happens, part of Earth's crust can be forced deep into its interior. The high heat and pressure cause the crust to melt and rise as magma. A final way that magma rises is over hot spots. Hot spots are exactly what they sound like--hot areas inside of Earth. These areas heat up magma. The magma becomes less dense. When it is less dense it rises. Each of the reasons for rising magma are a bit different, but each can form volcanoes.

Mountain Building

New mountains are built when rocks are pushed upwards by the movement of the giant rocky plates that make up the Earth's crust. The rocks are pushed upwards in two ways: FOLD mountains are formed when layers of rock become buckled, and BLOCK mountains are formed when giant lumps of rock rise or fall. Volcanic eruptions also create mountains. Many mountain ranges have been built up and eroded away since the Earth was formed.

The Andes

The Andes is the longest mountain range on land. It was formed along the western margin of South America, where two tectonic plates (rocky plates that make up the Earth's crust) collided. The mountains are still rising by about 10 cm (4 in) every century.

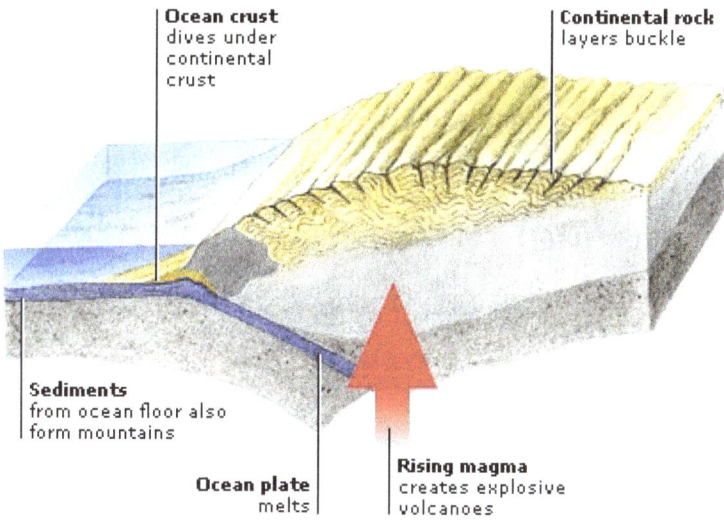

Ocean crust dives under continental crust

Continental rock layers buckle

Sediments from ocean floor also form mountains

Ocean plate melts

Rising magma creates explosive volcanoes

Fold mountains are pushed up at a boundary where two tectonic plates collide. The boundary between an ocean plate and a continental plate is called a subduction zone. Here, the thin ocean crust slides slowly under a thicker continental crust, making the rocks buckle and fold. The ocean plate also melts, creating magma (molten rock) that rises to form volcanoes.

World Mountain Ranges

The world's major mountain ranges, such as the Andes, the Himalayas, and the Alps, are situated along the boundaries where tectonic plates collide. These ranges formed in the last few hundred million years, so are they quite young. The map also shows thin lines of volcanoes that erupt from the ocean floor, forming chains of mountainous islands.

Himalayan Collision

The Himalayas is a range of fold mountains formed by the collision between India and the rest of Asia. When the two tectonic plates collided, the southern edge of Asia buckled. The Indian plate continues to slide under Asia and, to date, has uplifted Tibet to a height of over 5 km (3 miles).

Fold Formation

When layers of rock are pushed inwards from both ends, they crumple up into waves called folds. Rocks are too hard to be squashed into a smaller space. Instead they fold upwards and downwards. The immense forces that cause folding can crunch solid rocks into folds just a few metres across.

Folding Layers

The rocks that buckle to form fold mountains are made up of layers of sedimentary rocks and igneous rocks. When the layers are folded, the rocks on the outside of a fold are stretched and the rocks on the inside of a fold are squashed. The folding also makes the layers of rock slide over each other.

Block Mountains

Block mountains are mountains formed when layers of rock crack into giant blocks. Cracks in layers of rock are called faults. They form when the Earth's crust is stretched, squashed, or twisted. The blocks are free to slip up, down, or sideways, or to tip over. These movements are very slow, but over millions of years they form mountains thousands of metres high.

Weathering

Rocks, minerals, soils normally change their structure under the action or influence of certain environmental forces. Biological activity, extreme weather, and agents of erosion such as water, wind and ice are examples of environmental forces that influence the continuous breakdown, wearing away and loosening of rocks and soils. This is what is termed as weathering.

Weathering is thus the process where rocks or soils are dissolved or worn away into smaller and smaller pieces due to particular environmental factors such as the examples given above. In geological terms, weathering is defined as the disintegration of rocks influenced by animal and plant life, water, and the atmospheric forces in general.

Weathering is different from erosion. While erosion is the process by which soil and rock particles are worn away and moved elsewhere by wind, water or ice, weathering involves no moving agent of transport. It is the process of breakdown of rocks at the Earth's surface, either by extreme temperatures or rainwater or biological activity. It simply does not involve any movement of rock material.

Weathering processes are of three main types: mechanical, organic and chemical weathering.

Mechanical or Physical Weathering

Mechanical weathering is also known as physical weathering. Mechanical weathering is the physical breakdown of rocks into smaller and smaller pieces. One of the most common mechanical actions is frost shattering. It happens when water enters the pores and cracks of rocks, then freezes. Frost weathering, frost wedging, ice wedging or cry fracturing is the collective name for several processes where ice is present. These processes include frost shattering, frost-wedging and freeze-thaw weathering.

Once the frozen water is within the rocks, it expands by about 10% thereby opening the cracks a bit wider. The pressure acting within the rocks is estimated at 30,000 pounds per square inch at -7.6°F. Over time, this pressure alongside the changes in weather makes the rock split off, and bigger rocks are broken into smaller fragments.

Another type of mechanical weathering is called salt wedging. Winds, water waves, and rain also have an effect on rocks as they are physical forces that wear away rock particles, particularly over long periods of time. These forces are equally categorized under mechanical or physical weathering because they release their pressures on the rocks directly and indirectly which causes the rocks to fracture and disintegrate.

Mechanical/physical weathering is also caused by thermal stress which is the contraction and expansion effect on the rocks caused by changes in temperature. Due to uneven expansion and contraction, the rocks crack apart and disintegrate into smaller pieces.

Organic or Biological Weathering

Organic or biological weathering refers to the same thing. It is the disintegration of rocks as a result of the action by living organisms. Trees and other plants can wear away rocks since as they penetrate into the soil and as their roots get bigger, they exert pressure on rocks and makes the cracks wider and deeper. Eventually, the plants break the rocks apart. Some plants also grow within the fissures in the rocks which lead to widening of the fissures and then eventual disintegration.

Microscopic organisms like algae, moss, lichens and bacteria can grow on the surface of the rocks and produce chemicals that have the potential of breaking down the outer layer of the rock. They eat away the surface of the rocks. These microscopic organisms also bring about moist chemical micro-environments which encourage the chemical and physical breakdown of the rock surfaces. The amount of biological activity depends upon how much life is in that area. Burrowing animals such as moles, squirrels and rabbits can speed up the development of fissures.

Chemical Weathering

Chemical weathering happens when rocks are worn away by chemical changes. The natural chemical reactions within the rocks change the composition of the rocks over time. Because the chemical processes are gradual and ongoing, the mineralogy of rocks changes over time thus making them wear away, dissolve, and disintegrate.

The chemical transformations occur when water and oxygen interacts with minerals within the rocks to create different chemical reactions and compounds through processes such as hydrolysis and oxidation. As a result, in the process of new material formations, pores and fissures are created in the rocks thus enhancing the disintegration forces.

Rainwater can also at times become acid when it mixes with acidic depositions in the atmosphere. Acid depositions are created in the atmosphere as a consequence of fossil fuel combustion that releases oxides of nitrogen, sulfur and carbon.

The resultant acid water from precipitation – (acid rain) reacts with the rock's mineral particles producing new minerals and salts that can readily dissolve or wear away the rock grains. Chemical weathering mostly depends on the rock type and temperature. For instance, limestone is more prone to chemical erosion compared to granite. Higher temperatures increase the rate of chemical weathering.

Erosion and Deposition

Erosion is defined as the removal of soil, sediment, regolith, and rock fragments from the landscape. Most landscapes show obvious evidence of erosion. Erosion is responsible for the creation of hills and valleys. It removes sediments from areas that were once glaciated, shapes the shorelines of lakes and coastlines, and transports material downslope from elevated sites. In order for erosion to occur three processes must take place: detachment, entrainment and transport. Erosion also requires a medium to move material. Wind, water, and ice are the mediums primarily responsible for erosion. Finally, the process of erosion stops when the transported particles fall out of the transporting medium and settle on a surface. This process is called deposition. Figure below illustrates an area of Death Valley, California where the effects of erosion and deposition can be easily seen.

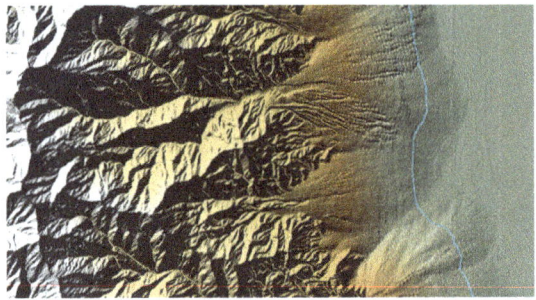

Digital Elevation Model

The following image was created from DEMs (Digital Elevation Model) for the following 1:24,000 scale topographic quadrangles: Telescope Peak, Hanaupah Canyon, and Badwater, California. To the left is the Panamint Mountain Range. To the right is Death Valley. Elevation spans from 3368 to -83 meters and generally decreases from left to right. The blue line represents an elevation of 0 meters. Large alluvial fans extending from a number of mountain valleys to the floor of Death Valley can be seen in the right side of the image. The sediments that make up these depositional features came from the weathering and erosion of bedrock in the mountains located on the left side of the image.

Energy of Erosion

The energy for erosion comes from several sources. Mountain building creates a disequilibrium within the Earth's landscape because of the creation of relief. Gravity acts to vertically move materials of higher relief to lower elevations to produce an equilibrium. Gravity also acts on the mediums of erosion to cause them to flow to base level.

Solar radiation and its influence on atmospheric processes is another source of energy for erosion. Rainwater has a kinetic energy imparted to it when it falls from the atmosphere. Snow has potential energy when it is deposited in higher elevations. This potential energy can be converted into the energy of motion when the snow is converted into flowing glacial ice. Likewise, the motion of air because of differences in atmospheric pressure can erode surface material when velocities are high enough to cause particle entrainment.

The Erosion Sequence

Erosion can be seen as a sequence of three events: detachment, entrainment, and transport. These three processes are often closely related and sometimes not easy distinguished between each other. A single particle may undergo detachment, entrainment, and transport many times.

Detachment

Erosion begins with the detachment of a particle from surrounding material. Sometimes detachment requires the breaking of bonds which hold particles together. Many different types of bonds exist each with different levels of particle cohesion. Some of the strongest bonds exist between the particles found within igneous rocks. In these materials, bonds are derived from the growth of mineral crystals during cooling. In sedimentary rocks, bonds are weaker and are mainly caused by the cementing effect of compounds such as iron oxides, silica, or calcium. The particles found in soils are held together by even weaker bonds which result from the cohesion effects of water and the electro-chemical bonds found in clay and particles of organic matter.

Physical, chemical, and biological weathering act to weaken the particle bonds found in rock materials. As a result, weathered materials are normally more susceptible than unaltered rock to the forces of detachment. The agents of erosion can also exert their own forces of detachment upon the surface rocks and soil through the following mechanisms:

Plucking: ice freezes onto the surface, particularly in cracks and crevices, and pulls fragments out from the surface of the rock.

Cavitation: intense erosion due to the surface collapse of air bubbles found in rapid flows of water. In the implosion of the bubble, a micro-jet of water is created that travels with high speeds and great pressure producing extreme stress on a very small area of a surface. Cavitation only occurs when water has a very high velocity, and therefore its effects in nature are limited to phenomenon like high waterfalls.

Raindrop impact: the force of a raindrop falling onto a soil or weathered rock surface is often sufficient to break weaker particle bonds. The amount of force exerted by a raindrop is a function of the terminal velocity and mass of the raindrop.

Abrasion: the excavation of surface particles by material carried by the erosion agent. The effectiveness of this process is related to the velocity of the moving particles, their mass, and their concentration at the eroding surface. Abrasion is very active in glaciers where the particles are firmly held by ice. Abrasion can also occur from the particles held in the erosional mediums of wind and water.

Entrainment

Entrainment is the process of particle lifting by the agent of erosion. In many circumstances, it is hard to distinguish between entrainment and detachment. There are several forces that provide particles with a resistance to this process. The most important force is frictional resistance. Frictional resistance develops from the interaction between the particle to its surroundings. A number of factors increase frictional resistance, including: gravity, particle slope angle relative to the flow direction of eroding medium, particle mass, and surface roughness.

Entrainment also has to overcome the resistance that occurs because of particle cohesive bonds. These bonds are weakened by weathering or forces created by the erosion agent (abrasion, plucking, raindrop impact, and cavitation).

Entrainment Forces

The main force reponsible for entrainment is fluid drag. The strength of fluid drag varies with the mass of the eroding medium (water is 9000 times more dense than air) and its velocity. Fluid drag causes the particle to move because of horizontal force and vertical lift. Within a medium of erosion, both of these forces are controlled by velocity. Horizontal force occurs from the push of the agent against the particle. If this push is sufficient to overcome friction and the resistance of cohesive bonds, the particle moves horizontally. The vertical lift is produced by turbulence or eddies within the flow that push the particle upward. Once the particle is lifted the only force resisting its transport is gravity as the forces of friction, slope angle, and cohesion are now non-existent. The particle can also be transported at velocities lower than the entrainment velocities because of the reduction in forces acting on it.

Many hydrologists and geomorphologists require a mathematical model to predict levels of entrainment, especially in stream environments. In these highly generalized models, the level of particle entrainment is relative to particle size and the velocity of the medium of erosion. These quantitative models can be represented graphically. On these graphs, the x-axis represents the log of particle diameter, and the y-axis the log of velocity. The relationship between these two variables to the entrainment of particles is described by a curve, and not by a straight line.

The critical entrainment velocity curve suggests that particles below a certain size are just as resistant to entrainment as particles with larger sizes and masses (See Figure). Fine silt and clay particles tend to have higher resistance to entrainment because of the strong cohesive bonds between particles. These forces are far stronger than the forces of friction and gravity.

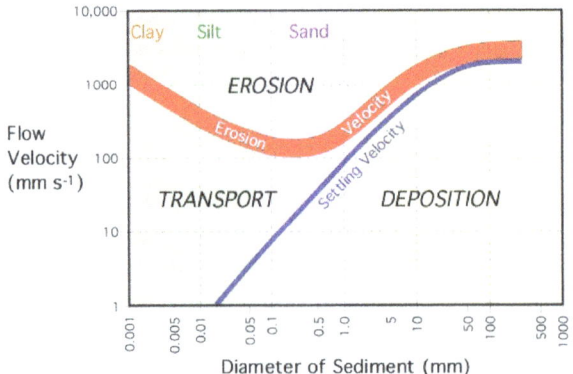

Figure: This graph describes the relationship between stream flow velocity
and particle erosion, transport, and deposition.

The curved line labeled "erosion velocity" describes the velocity required to entrain particles from the stream's bed and banks. The erosion velocity curve is drawn as a thick line because the erosion particles tends to be influenced by a variety of factors that changes from stream to stream. Also, note that the entrainment of silt and clay needs greater velocities then larger sand particles. This situation occurs because silt and clay have the ability to form cohesive bounds between particles. Because of the bonding, greater flow velocities are required to break the bonds and move these particles. The graph also indicates that the transport of particles requires lower flow velocities then erosion. This is especially true of silt and clay particles. Finally, the line labeled "settling velocity" shows at what velocity certain sized particles fall out of transport and are deposited.

Transport

Once a particle is entrained, it tends to move as long as the velocity of the medium is high enough to transport the particle horizontally. Within the medium, transport can occur in four different ways:

- Suspension is where the particles are carried by the medium without touching the surface of their origin. This can occur in air, water, and ice.

- Saltation is where the particle moves from the surface to the medium in quick continuous repeated cycles. The action of returning to the surface usually has enough force to cause the entrainment of new particles. This process is only active in air and water.

- Traction is the movement of particles by rolling, sliding, and shuffling along the eroded surface. This occurs in all erosional mediums.

- Solution is a transport mechanism that occurs only in aqueous environments. Solution involves the eroded material being dissolve and carried along in water as individual ions.

Particle weight, size, shape, surface configuration, and medium type are the main factors that determine which of these processes operate.

Deposition

The erosional transport of material through the landscape is rarely continuous. Instead, we find that particles may undergo repeated cycles of entrainment, transport, and deposition. Transport depends on an appropriate balance of forces within the transporting medium. A reduction in the velocity of the medium, or an increase in the resistance of the particles may upset this balance and cause deposition. Reductions in competence can occur in a variety of ways. Velocity can be reduced locally by the sheltering effect of large rocks, hills, stands of vegetation or other obstructions. Normally, competence changes occur because of large scale reductions in the velocity of flowing medium. For wind, reductions in velocity can be related to variations in spatial heating and cooling which create pressure gradients and wind. In water, lower velocities can be caused by reductions in discharge or a change in the grade of the stream. Glacial flows of ice can become slower if precipitation input is reduced or when the ice encounters melting. Deposition can also be caused by particle precipitation and flocculation. Both of these processes are active only in water. Precipitation is a process where dissolved ions become solid because of changes in the temperature or chemistry of the water. Flocculation is a chemical process where salt causes the aggregation of minute clay particles into larger masses that are too heavy to remain suspended.

References

- Formation-types-and-examples-of-igneous-rocks: eartheclipse.com, Retrieved 31 May 2018

- Baker, Victor R.; Nummedal, Dag, eds. (1978). The Channeled Scabland: A Guide to the Geomorphology of the Columbia Basin, Washington. Washington, D.C.: Planetary Geology Program, Office of Space Science, National Aeoronautics and Space Administration. pp. 173–177. ISBN 0-88192-590-X

- Metamorphic-rocks, local-rocks: geologyglasgow.org.uk, Retrieved 19 June 2018

- Earthquakes-definition-causes-measures-and-other-details-with-diagram-31852: yourarticlelibrary.com, Retrieved 21 June 2018

- Wilkinson, Bruce H.; McElroy, Brandon J.; Kesler, Stephen E.; Peters, Shanan E.; Rothman, Edward D. (2008). "Global geologic maps are tectonic speedometers—Rates of rock cycling from area-age frequencies". Geological Society of America Bulletin. 121: 760–779. doi:10.1130/B26457.1

- Mountain-building, science: factmonster.com, Retrieved 31 March 2018

- Different-types-of-weathering, geology: eartheclipse.com, Retrieved 19 July 2018

The Hydrosphere

The hydrosphere is the total sum of all the water found in Earth's oceans, seas, rivers, lakes, streams and ponds. The chapter closely examines the key aspects of the hydrosphere through the elucidation of the physical properties of water, water cycle, water distribution on Earth, physical oceanography, ocean current, etc. for a comprehensive understanding.

Physical Properties of Water

Water has a high specific heat, i.e. it needs a lot of heat to heat up and takes long to lose the stored heat and get cold. This is why it is used in cooling systems (for instance in car radiators or to cool industrial equipment). And this is also why in coastal (or lake) regions the temperature of the air is milder: in these areas, as seasons change, the temperature of the water 'mitigates' the temperature of the air, since it decreases or increases more slowly than that of the air. Water has a high surface tension: that means that, once poured on a smooth surface, it tends to form spherical drops instead of expanding into a thin film. Without gravity, a drop of water would be perfectly spherical. Surface tension allows plants to absorb the water contained in the soil through their roots. And it is surface tension, again, that makes blood, which is largely composed of water molecules, flow through the blood system of our body.

In addition, water can normally be found in a liquid state, but can easily become solid or gaseous. Pure water goes from liquid to solid, i.e. freezes, at 0 degrees centigrade, while at sea level it boils at 100°C (the higher the level, the lower the temperature at which water starts boiling). The water boiling and freezing values are taken as a reference point to calibrate thermometers: in centigrade scales, 0° on the centigrade scale is the freezing point and 100° is the boiling point.

When freezing, water expands, i.e. its density decreases while its volume remains the same: this is why ice floats on the water or a bottle filled with water and placed in a freezer breaks up.

Water is a special natural resource since it is the only one on earth to be found in all of the three physical states depending on the surrounding temperature: liquid, solid (ice) and gaseous (water vapour).

The whole of the processes that make water leave the oceans, get into the atmosphere, reach the emerged lands and flow back to the oceans later on is called hydrological cycle and is fuelled by the energy of the Sun.

Physical Properties

Water is the chemical substance with chemical formula H_2O; one molecule of water has two hydrogen atoms covalently bonded to a single oxygen atom.[Water is a tasteless, odorless liquid at ambient temperature and pressure, and appears colorless in small quantities, although it has its own intrinsic very light blue hue. Ice also appears colorless, and water vapor is essentially invisible as a gas.

Unlike other analogous hydrides of the oxygen family, water is primarily a liquid under standard conditions due to hydrogen bonding. The molecules of water are constantly moving in relation to each other, and the hydrogen bonds are continually breaking and reforming at timescales faster than 200 femtoseconds (2×10^{-13} seconds). However, these bonds are strong enough to create many of the peculiar properties of water, some of which make it integral to life.

Water, Ice and Vapor

Within the Earth's atmosphere and surface, the liquid phase is the most common and is the form that is generally denoted by the word "water". The solid phase of water is known as ice and commonly takes the structure of hard, amalgamated crystals, such as ice cubes, or loosely accumulated granular crystals, like snow. Aside from common hexagonal crystalline ice, other crystalline and amorphous phases of ice are known. The gaseous phase of water is known as water vapor (or steam). Visible steam and clouds are formed from minute droplets of water suspended in the air.

Water also forms a supercritical fluid. The critical temperature is 647 K and the critical pressure is 22.064 MPa. In nature this only rarely occurs in extremely hostile conditions. A likely example of naturally occurring supercritical water is in the hottest parts of deep water hydrothermal vents, in which water is heated to the critical temperature by volcanic plumes and the critical pressure is caused by the weight of the ocean at the extreme depths where the vents are located. This pressure is reached at a depth of about 2200 meters: much less than the mean depth of the ocean (3800 meters).

Heat Capacity and Heats of Vaporization and Fusion

Water has a very high specific heat capacity of 4.1814 J/(g·K) at 25°C – the second highest among all the heteroatomic species (after ammonia), as well as a high heat of vaporization (40.65 kJ/mol or 2257 kJ/kg at the normal boiling point), both of which are a result of the extensive hydrogen bonding between its molecules. These two unusual properties allow water to moderate Earth's climate by buffering large fluctuations in temperature. Most of the additional energy stored in the climate system since 1970 has accumulated in the oceans.

The specific enthalpy of fusion (more commonly known as latent heat) of water is 333.55 kJ/kg at 0°C: the same amount of energy is required to melt ice as to warm ice from −160°C up to its melting point or to heat the same amount of water by about 80 °C. Of common substances, only that of ammonia is higher. This property confers resistance to melting on the ice of glaciers and drift ice. Before and since the advent of mechanical refrigeration, ice was and still is in common use for retarding food spoilage.

The specific heat capacity of ice at −10°C is 2.03 J/(g·K) and the heat capacity of steam at 100°C is 2.08 J/(g·K).

The density of water is about 1 gram per cubic centimetre (62 lb/cu ft): this relationship was originally used to define the gram. The density varies with temperature, but not linearly: as the temperature increases, the density rises to a peak at 3.98°C (39.16 °F) and then decreases. This unusual negative thermal expansion below 4°C (39 °F) is also observed in molten silica. Regular, hexagonal ice is also less dense than liquid water—upon freezing, the density of water decreases by about 9%.

These effects are due to the reduction of thermal motion with cooling, which allows water molecules to form more hydrogen bonds that prevent the molecules from coming close to each other. While below 4°C the breakage of hydrogen bonds due to heating allows water molecules to pack closer despite the increase in the thermal motion (which tends to expand a liquid), above 4°C water expands as the temperature increases. Water near the boiling point is about 4% less dense than water at 4°C (39 °F).

Under increasing pressure, ice undergoes a number of transitions to other polymorphs with higher density than liquid water, such as ice II, ice III, high-density amorphous ice (HDA), and very-high-density amorphous ice (VHDA).

The unusual density curve and lower density of ice than of water is vital to life—if water were most dense at the freezing point, then in winter the very cold water at the surface of lakes and other water bodies would sink, the lake could freeze from the bottom up, and all life in them would be killed. Furthermore, given that water is a good thermal insulator (due to its heat capacity), some frozen lakes might not completely thaw in summer. The layer of ice that floats on top insulates the water below. Water at about 4°C (39 °F) also sinks to the bottom, thus keeping the temperature of the water at the bottom constant.

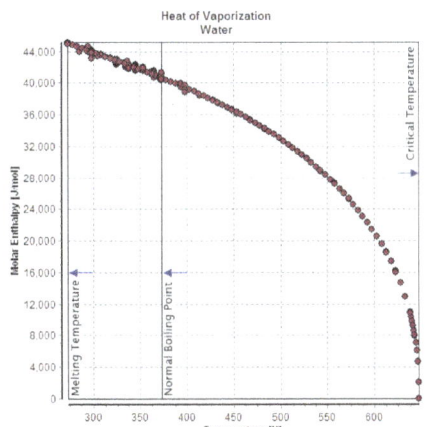

Heat of vaporization of water from melting to critical temperature

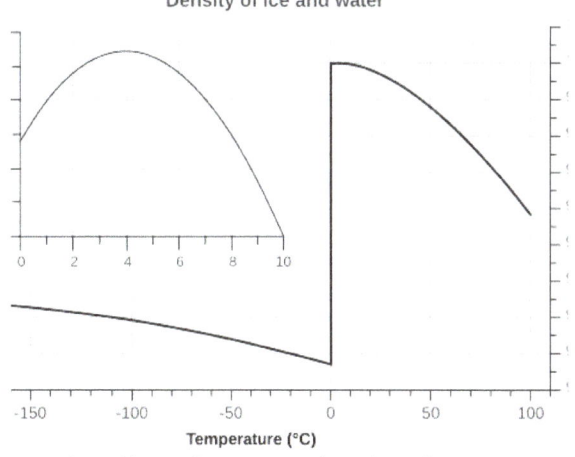

Density of ice and water as a function of temperature

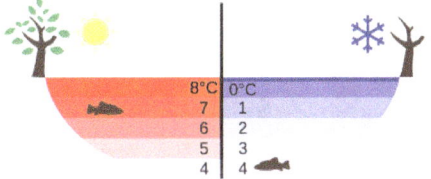

Temperature distribution in a lake in summer and winter

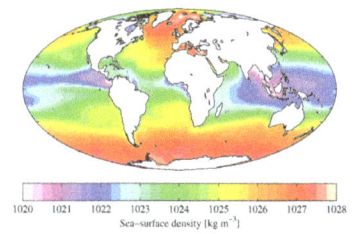

WOA surface density

Density of Saltwater and Ice

The density of salt water depends on the dissolved salt content as well as the temperature. Ice still floats in the oceans, otherwise they would freeze from the bottom up. However, the salt content of oceans lowers the freezing point by about 1.9 °C and lowers the temperature of the density maximum of water to the former freezing point at 0°C. This is why, in ocean water, the downward convection of colder water is not blocked by an expansion of water as it becomes colder near the freezing point. The oceans' cold water near the freezing point continues to sink. So creatures that live at the bottom of cold oceans like the Arctic Ocean generally live in water 4° C colder than at the bottom of frozen-over fresh water lakes and rivers.

As the surface of salt water begins to freeze (at −1.9° C for normal salinity seawater, 3.5%) the ice that forms is essentially salt-free, with about the same density as freshwater ice. This ice floats on the surface, and the salt that is "frozen out" adds to the salinity and density of the sea water just below it, in a process known as *brine rejection*. This denser salt water sinks by convection and the replacing seawater is subject to the same process. This produces essentially freshwater ice at −1.9 °C on the surface. The increased density of the sea water beneath the forming ice causes it to sink towards the bottom. On a large scale, the process of brine rejection and sinking cold salty water results in ocean currents forming to transport such water away from the Poles, leading to a global system of currents called the thermohaline circulation.

Miscibility and Condensation

Water is miscible with many liquids, including ethanol in all proportions. Water and most oils are immiscible usually forming layers according to increasing density from the top. This can be predicted by comparing the polarity. Water being a relatively polar compound will tend to be miscible with liquids of high polarity such as ethanol and acetone, whereas compounds with low polarity will tend to be immiscible and poorly soluble such as with hydrocarbons.

As a gas, water vapor is completely miscible with air. On the other hand, the maximum water vapor pressure that is thermodynamically stable with the liquid (or solid) at a given temperature is relatively low compared with total atmospheric pressure. For example, if the vapor's partial pressure is 2% of atmospheric pressure and the air is cooled from 25 °C, starting at about 22 °C water will start to condense, defining the dew point, and creating fog or dew. The reverse process accounts for the fog burning off in the morning. If the humidity is increased at room temperature, for example, by running a hot shower or a bath, and the temperature stays about the same, the vapor soon reaches the pressure for phase change, and then condenses out as minute water droplets, commonly referred to as steam.

A saturated gas or one with 100% relative humidity is when the vapor pressure of water in the air is at equilibrium with vapor pressure due to (liquid) water; water (or ice, if cool enough) will fail to lose mass through evaporation when exposed to saturated air. Because the amount of water vapor in air is small, relative humidity, the ratio of the partial pressure due to the water vapor to the saturated partial vapor pressure, is much more useful. Vapor pressure above 100% relative humidity is called super-saturated and can occur if air is rapidly cooled, for example, by rising suddenly in an updraft.

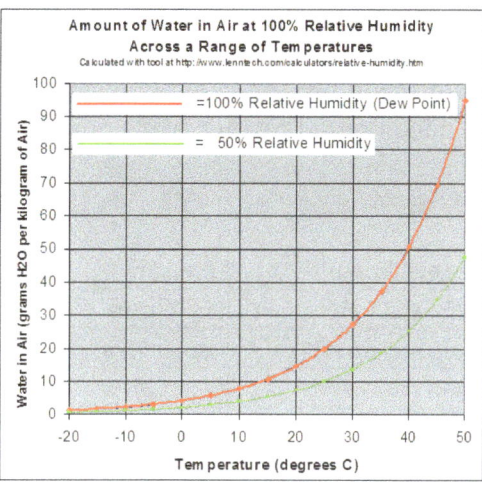

Red line shows saturation

Vapor Pressure

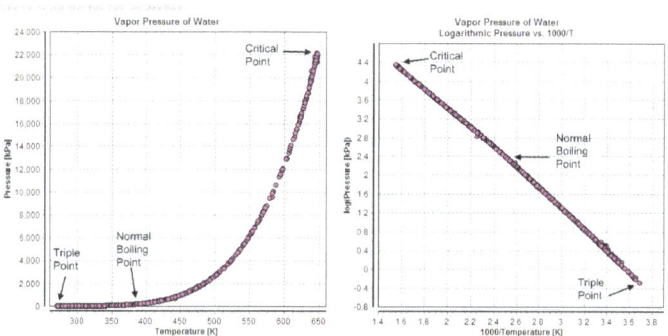

Vapor pressure diagrams of water

Compressibility

The compressibility of water is a function of pressure and temperature. At 0 °C, at the limit of zero pressure, the compressibility is 5.1×10^{-10} Pa^{-1}. At the zero-pressure limit, the compressibility reaches a minimum of 4.4×10^{-10} Pa^{-1} around 45 °C before increasing again with increasing temperature. As the pressure is increased, the compressibility decreases, being 3.9×10^{-10} Pa^{-1} at 0 °C and 100 megapascals (1,000 bar).

The bulk modulus of water is about 2.2 GPa. The low compressibility of non-gases, and of water in particular, leads to their often being assumed as incompressible. The low compressibility of water means that even in the deep oceans at 4 km depth, where pressures are 40 MPa, there is only a 1.8% decrease in volume.

Triple Point

The temperature and pressure at which ordinary solid, liquid, and gaseous water coexist in equilibrium is a triple point of water. This point has been used to define the base unit of temperature, the kelvin, since 1954, and is thus set as having a temperature of 273.16 kelvin.

Due the existence of many polymorphs (forms) of ice, water has other triple points, which have either three polymorphs of ice or two polymorphs of ice and liquid in equilibrium. Gustav Heinrich Johann Apollon Tammann in Göttingen produced data on several other triple points in the early 20th century. Kamb and others documented further triple points in the 1960s.

The Various Triple Points of Water

Phases in stable equilibrium	Pressure	Temperature
liquid water, ice I_h, and water vapor	611.657 Pa	273.16 K (0.01 °C)
liquid water, ice I_h, and ice III	209.9 MPa	251 K (−22 °C)
liquid water, ice III, and ice V	350.1 MPa	−17.0 °C
liquid water, ice V, and ice VI	632.4 MPa	0.16 °C
ice I_h, Ice II, and ice III	213 MPa	−35 °C
ice II, ice III, and ice V	344 MPa	−24 °C
ice II, ice V, and ice VI	626 MPa	−70 °C

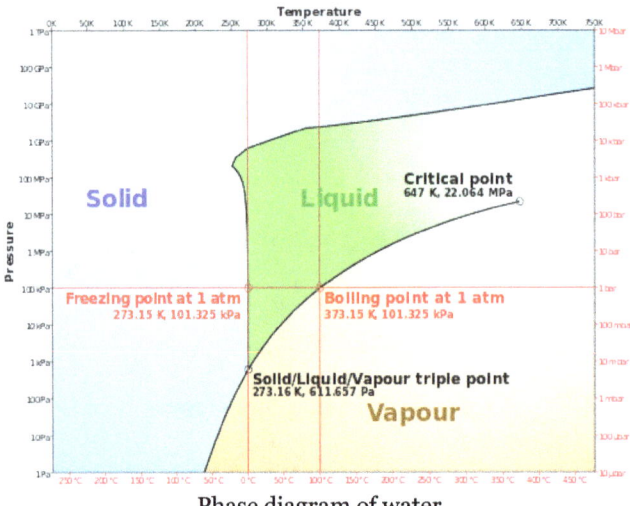

Phase diagram of water

Melting Point

The melting point of ice is 0 °C (32 °F; 273 K) at standard pressure; however, pure liquid water can be supercooled well below that temperature without freezing if the liquid is not mechanically disturbed. It can remain in a fluid state down to its homogeneous nucleation point of about 231 K (−42 °C; −44 °F).[The melting point of ordinary hexagonal ice falls slightly under moderately high pressures, by 0.0073 °C (0.0131 °F)/atm or about 0.5 °C (0.90 °F)/70 atm as the stabilization energy of hydrogen bonding is exceeded by intermolecular repulsion, but as ice transforms into its allotropes above 209.9 MPa (2,072 atm), the melting point increases markedly with pressure, i.e., reaching 355 K (82 °C) at 2.216 GPa (21,870 atm) (triple point of Ice VII).

Electrical Properties

Electrical Conductivity

Pure water containing no exogenous ions is an excellent insulator, but not even "deionized" water is completely free of ions. Water undergoes auto-ionization in the liquid state, when two water molecules form one hydroxide anion OH^-) and one hydronium cation H_3O^+).

Because water is such a good solvent, it almost always has some solute dissolved in it, often a salt. If water has even a tiny amount of such an impurity, then it can conduct electricity far more readily.

It is known that the theoretical maximum electrical resistivity for water is approximately 18.2 $M\Omega \cdot cm$ (182 $k\Omega \cdot m$) at 25 °C. This figure agrees well with what is typically seen on reverse osmosis, ultra-filtered and deionized ultra-pure water systems used, for instance, in semiconductor manufacturing plants. A salt or acid contaminant level exceeding even 100 parts per trillion (ppt) in otherwise ultra-pure water begins to noticeably lower its resistivity by up to several $k\Omega \cdot m$.

In pure water, sensitive equipment can detect a very slight electrical conductivity of 0.05501 ± 0.0001 µS/cm at 25.00 °C. Water can also be electrolyzed into oxygen and hydrogen gases but in the absence of dissolved ions this is a very slow process, as very little current is conducted. In ice, the primary charge carriers are protons. Ice was previously thought to have a small but measurable conductivity of 1×10^{-10} S/cm, but this conductivity is now thought to be almost entirely from surface defects, and without those, ice is an insulator with an immeasurably small conductivity.

Polarity and Hydrogen Bonding

An important feature of water is its polar nature. The structure has a bent molecular geometry for the two hydrogens from the oxygen vertex. The oxygen atom also has two lone pairs of electrons. One effect usually ascribed to the lone pairs is that the H–O–H gas phase bend angle is 104.48°,which is smaller than the typical tetrahedral angle of 109.47°. The lone pairs are closer to the oxygen atom than the electrons sigma bonded to the hydrogens, so they require more space. The increased repulsion of the lone pairs forces the O–H bonds closer to each other.

Another consequence of its structure is that water is a polar molecule. Due to the difference in electronegativity, a bond dipole moment points from each H to the O, making the oxygen partially negative and each hydrogen partially positive. A large molecular dipole, points from a region between the two hydrogen atoms to the oxygen atom. The charge differences cause water molecules to aggregate (the relatively positive areas being attracted to the relatively negative areas). This attraction, hydrogen bonding, explains many of the properties of water, such as its solvent properties.

Although hydrogen bonding is a relatively weak attraction compared to the covalent bonds within the water molecule itself, it is responsible for a number of water's physical properties. These properties include its relatively high melting and boiling point temperatures: more energy is required to break the hydrogen bonds between water molecules. In contrast, hydrogen sulfide (H_2S), has much weaker hydrogen bonding due to sulfur's lower electronegativity. H_2S is a gas at room temperature, in spite of hydrogen sulfide having nearly twice the molar mass of water. The extra bonding between water molecules also gives liquid water a large specific heat capacity. This high heat capacity makes water a good heat storage medium (coolant) and heat shield.

A diagram showing the partial charges on
the atoms in a water molecule

Cohesion and Adhesion

Water molecules stay close to each other (cohesion), due to the collective action of hydrogen bonds between water molecules. These hydrogen bonds are constantly breaking, with new bonds being formed with different water molecules; but at any given time in a sample of liquid water, a large portion of the molecules are held together by such bonds.

Dew drops adhering to a spider web

Water also has high adhesion properties because of its polar nature. On extremely clean/smooth glass the water may form a thin film because the molecular forces between glass and water molecules (adhesive forces) are stronger than the cohesive forces. In biological cells and organelles, water is in contact with membrane and protein surfaces that are hydrophilic; that is, surfaces that have a strong attraction to water. Irving Langmuir observed a strong repulsive force between hydrophilic surfaces. To dehydrate hydrophilic surfaces—to remove the strongly held layers of water of hydration—requires doing substantial work against these forces, called hydration forces. These forces are very large but decrease rapidly over a nanometer or less. They are important in biology, particularly when cells are dehydrated by exposure to dry atmospheres or to extracellular freezing.

Surface Tension

This paper clip is under the water level, which has risen gently and smoothly. Surface tension prevents the clip from sumerging and the water from overflowing the glass edges

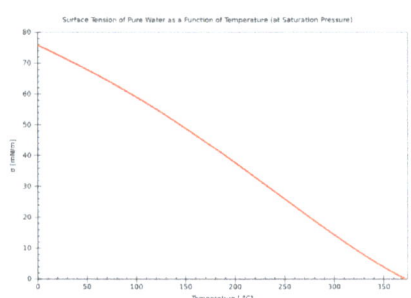

Temperature dependence of the
surface tension of pure water

Water has an unusually high surface tension of 71.99 mN/m at 25 °C which is caused by the strength of the hydrogen bonding between water molecules. This allows insects to walk on water.

Capillary Action

Because water has strong cohesive and adhesive forces, it exhibits capillary action. Strong cohesion from hydrogen bonding and adhesion allows trees to transport water more than 100 m upward.

Water as a Solvent

Water is an excellent solvent due to its high dielectric constant. Substances that mix well and dissolve in water are known as hydrophilic ("water-loving") substances, while those that do not mix well with water are known as hydrophobic ("water-fearing") substances. The ability of a substance to dissolve in water is determined by whether or not the substance can match or better the strong attractive forces that water molecules generate between other water molecules. If a substance has properties that do not allow it to overcome these strong intermolecular forces, the molecules are precipitated out from the water. Contrary to the common misconception, water and hydrophobic substances do not "repel", and the hydration of a hydrophobic surface is energetically, but not entropically, favorable.

When an ionic or polar compound enters water, it is surrounded by water molecules (hydration). The relatively small size of water molecules (~ 3 angstroms) allows many water molecules to surround one molecule of solute. The partially negative dipole ends of the water are attracted to positively charged components of the solute, and vice versa for the positive dipole ends.

In general, ionic and polar substances such as acids, alcohols, and salts are relatively soluble in water, and non-polar substances such as fats and oils are not. Non-polar molecules stay together in water because it is energetically more favorable for the water molecules to hydrogen bond to each other than to engage in van der Waals interactions with non-polar molecules.

An example of an ionic solute is table salt; the sodium chloride, NaCl, separates into Na^+ cations and Cl^- anions, each being surrounded by water molecules. The ions are then easily transported away from their crystalline lattice into solution. An example of a nonionic solute is table sugar. The water dipoles make hydrogen bonds with the polar regions of the sugar molecule (OH groups) and allow it to be carried away into solution.

Presence of colloidal calcium carbonate from high concentrations of
dissolved lime turns the water of Havasu Falls turquoise

Quantum Tunneling

The quantum tunneling dynamics in water was reported as early as 1992. At that time it was known that there are motions which destroy and regenerate the weak hydrogen bond by internal rotations of the substituent water monomers. On 18 March 2016, it was reported that the hydrogen bond can be broken by quantum tunneling in the water hexamer. Unlike previously reported tunneling motions in water, this involved the concerted breaking of two hydrogen bonds. Later in the same year, the discovery of the quantum tunneling of water molecules was reported.

Electromagnetic Absorption

Water is relatively transparent to visible light, near ultraviolet light, and far-red light, but it absorbs most ultraviolet light, infrared light, and microwaves. Most photoreceptors and photosynthetic pigments utilize the portion of the light spectrum that is transmitted well through water. Microwave ovens take advantage of water's opacity to microwave radiation to heat the water inside of foods. Water's light blue colour is caused by weak absorption in the red part of the visible spectrum.

Water Cycle

The water cycle, or the hydrologic cycle, is the continuous circulation of water within the Earth's hydrosphere. It involves the movement of water into and out of various reservoirs, including the atmosphere, land, surface water, and groundwater. This cycle is driven by radiation from the Sun. The movement of water within the water cycle is the subject of the field of hydrology.

The water moves from one reservoir to another, such as from river to ocean, or from the ocean to the atmosphere, by the physical processes of evaporation, condensation, precipitation, infiltration, runoff, and subsurface flow. In so doing, the water goes through different phases: liquid, solid, and gas.

The water cycle also involves the exchange of heat energy, which leads to temperature changes. For instance, in the process of evaporation, water takes up energy from the surroundings and cools the environment. Conversely, in the process of condensation, water releases energy to its surroundings, warming the environment.

The water cycle figures significantly in the maintenance of life and ecosystems on Earth. Even as water in each reservoir plays an important role, the water cycle brings added significance to the presence of water on our planet. By transferring water from one reservoir to another, the water cycle purifies water, replenishes the land with freshwater, and transports minerals to different parts of the globe. It is also involved in reshaping the geological features of the Earth, through such processes as erosion and sedimentation. In addition, as the water cycle involves heat exchange, it exerts an influence on climate as well.

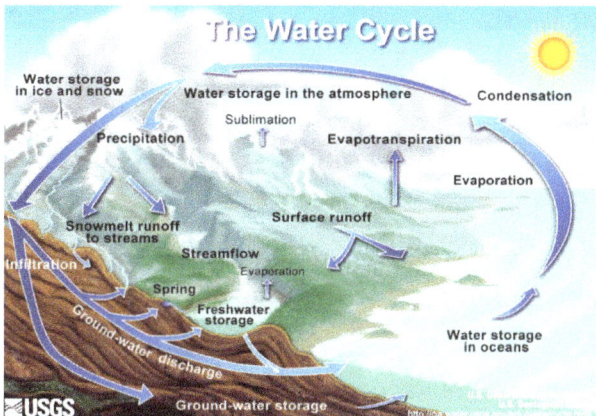

The movement of water around, over, and through the Earth is called the water cycle

Movement of Water within the Water Cycle

There is no definable start or finish to the water cycle. Water molecules move continuously among different compartments, or reservoirs, of the Earth's hydrosphere, by different physical processes. Water evaporates from the oceans, forms clouds, which precipitate and the water falls back to Earth. However, water does not necessarily cycle through each compartment in order. Before reaching the ocean, water may have evaporated, condensed, precipitated, and become runoff multiple times.

Explanation of the Water Cycle

This cycle takes place in just about few main parts:

Evaporation and Transpiration

Evaporation is the point at which the sun warms up water in streams or lakes or the sea and transforms it into vapor or steam. The water vapor or steam leaves the stream, lake or sea and goes into the air.

Everybody knows about the procedure of evaporation. Assume that you spill a teaspoon of water on the kitchen table. In the event that you return a couple of hours later, the water will have vanished. It has changed from fluid water into water vapor, or evaporated.

Importance of Evaporation

Evaporation is a vital piece of the water cycle. Heat from the sun, or solar energy, controls the evaporation process. It absorbs moisture from soil in a garden, and additionally the greatest seas and lakes. The water level will decline as it is bare to the heat of the sun.

Factors that Affect Evaporation

Some liquids evaporate more rapidly than others. There are many factors that influence the evaporation rate.

- In the event that the air is already congested, or saturate, with different substances, there won't be sufficient room noticeable all around for liquid to evaporate rapidly. At the point when the humidity is 100 percent, the air is saturated with water. No more water can evaporate.

- Air pressure likewise influences evaporation. Whenever the air pressure is high on the surface of a waterway, at that point the water won't evaporate efficiently. The pressure pushing down on the water makes it hard for water to escape into the air as vapor. Storms are regularly high-pressure systems that avoid evaporation.

- Temperature, obviously, influences how rapidly evaporation happens. Boiling heated water will evaporate rapidly as steam.

Transpiration is the process by which plants lose water out of their leaves. Transpiration provides evaporation somewhat of a turn in recovering the water vapor up into the air.

Plant transpiration is basically an undetectable process, since the water is evaporating from the leaf surfaces, you don't simply go out and see the leaves "sweating". Because you can't see the water doesn't mean it isn't being put into the air. Amid a growing season, a leaf will unfold ordinarily more water than its own particular weight. A section of land of corn emits around 3,000-4,000 gallons (11,400-15,100 liters) of water every day, and a vast oak tree can happen 40,000 gallons (151,000 liters) every year.

Sublimation

Sublimation describes the process of snow and ice changing into water vapor without first melting into water. Sublimation is a common way for snow to disappear in certain climates.

It is not easy to actually see sublimation happen, at least not with ice. One way to see the results of sublimation is to hang a wet shirt outside on a below-freezing day. Eventually the ice in the shirt will disappear. Actually, the best way to visualize sublimation is to not use water at all, but to use carbon dioxide instead, as this picture shows."Dry ice" is solid, frozen carbon dioxide, which sublimates, or turns to gas, at the temperature -78.5 °C (-109.3°F). The fog you see in the picture is a mixture of cold carbon dioxide gas and cold, humid air, created as the dry ice sublimates.

Sublimation occurs more readily when certain weather conditions are present, such as low relative humidity and dry winds. It also occurs more at higher altitudes, where the air pressure is less than at lower altitudes.

Energy, such as strong sunlight, is also needed. If I was to pick one place on Earth where sublimation happens a lot. Low temperatures, strong winds, intense sunlight, very low air pressure - just what is needed for sublimation to occur.

Condensation

Water vapor in the air gets cold and changes once more into liquid, shaping mists or clouds. This is called condensation.

You can see a similar kind of thing at home. Set out a glass of cold water on a hot day and observe what happens. Water forms outside of the glass. That water didn't some way or another spill through the glass! It really originated from the air. Water vapor in the warm air, transforms once more into liquid when it touches the cold glass.

Importance of Condensation

Condensation is essential to the water cycle since it is in charge of the development of clouds. These clouds may deliver precipitation, that will be tackled later, which is the essential course for water to come back to the Earth's surface inside the water cycle. Condensation is the exact opposite of evaporation.

Clouds form when water vapor condenses around little particles, similar to bits of dust or smoke noticeable all around. Relying upon the amount of the drops, these particles might possibly be noticeable. Even on a clear, cloudless day, water vapor is constantly present in the environment, yet it varies in numbers. We know it is visible on an extremely humid day; it frequently feels like we have to swim through the air! Fog is condensation close to the ground.

Causes of Condensation

Like evaporation, condensation happens as a major aspect of the water cycle. Water molecules that have moved upward through evaporation in the end meet the cooler air at more elevated amounts of the climate. Water vapor in the warm, moist air condenses, shaping bigger beads of water that will inevitably be noticeable as clouds.

The reason is the adjustment in temperature. The cooler air can't keep water particles isolated, so they join again to form droplets. Condensation is happening regardless of the possibility that clouds are not visible. As more water vapor condenses, clouds normally start to shape. Precipitation takes after, and the water cycle starts once more.

Precipitation

The next phase of the water cycle is Precipitation, happens when so much water has condensed that the air can't hold it any longer. The clouds get substantial and water falls back to the earth in the form of rain, hail, sleet, snow or freezing rain.

Clouds are necessary for precipitation because the raindrops are the drops of the clouds that have sufficiently condensed water to start falling. The cloud particles don't have enough mass to fall, however as condensation keeps on adding water to those particles, gravity in the long run pulls them towards the Earth as precipitation.

Around 505,000 km3 (121,000 cu mi) of water falls as precipitation every year, 398,000 km3 (95,000 cu mi) of it over the oceans. The rain on land contains 107,000 km3 (26,000 cu mi) of water every year and a snowing just 1,000 km3 (240 cu mi).

Importance of Precipitation

Precipitation is expected to recharge water to the earth. Without precipitation, this planet would be a huge desert. The number and span of precipitation occurrence influence both water level and water quality inside an estuary.

Precipitation supplies freshwater to an estuary, which is an imperative wellspring of dissolved oxygen and supplements. Droughts bring down the freshwater contribution to estuaries and the water levels of inland lakes. Lake levels impact water waste and flow patterns in freshwater estuaries.

Measurement of Precipitation

Precipitation is normally described in millimeters or inches of liquid precipitation. This number is normally included over a specific timeframe, for example, inches every day.

Factors that Affect Precipitation

Massive precipitation happens close to the equator and reduces with the expansion in the latitude like towards Polar Regions. Primary source of moisture for precipitation is evaporation from seas. Hence, precipitation has a tendency to be heavier close to coastlines.

Since lifting of air masses is the reason for all precipitation, amount and recurrence of rain is generally more on windward side of the mountain. As downslope movement of air brings about reduction in humidity, in this manner the inverse sides of barriers commonly experience moderately light precipitation. High amount of precipitation is accounted for at higher elevations.

- Prevailing Winds- winds move moist air over land

- Presents of mountains- mountain range can change the path of prevailing winds and impact where precipitation falls

- Seasons- sea and land breezes that change directions with the season; is known as monsoon.

Infiltration

Anywhere in the world, a portion of the water that falls as rain and snow infiltrates into the sub-surface soil and rock. How much infiltrates depends greatly on a number of factors. Infiltration of precipitation falling on the ice cap of Greenland might be very small.

Infiltration is the downward movement of water from the land surface into soil or porous rock. Some water that infiltrates will remain in the shallow soil layer, where it will gradually move vertically and horizontally through the soil and subsurface material. Eventually it might enter a stream by seepage into the stream bank. Some of the water may infiltrate deeper, recharging groundwater aquifers.

If the aquifers are shallow or porous enough to allow water to move freely through it, people can drill wells into the aquifer and use the water for their purposes. Water may travel long distances or remain in groundwater storage for long periods before returning to the surface or seeping into other water bodies, such as streams and the oceans.

In places where the water table (the top of the saturated zone) is close to the land surface and where the water can move through the aquifer at a high rate, aquifers can be replenished artificially.

Collection

At the point when water falls back to earth as precipitation, it might fall back in the seas, lakes or rivers or it might wind up ashore. When it winds up ashore, it will either splash into the earth or turn out to be a piece of the "ground water" that plants and creatures use to drink or it might keep running over the dirt and gather in the seas, lakes or streams where the cycle starts from the very beginning once more.

Human Activities that Change the Water Cycle Include

- Agriculture

- Industry

- Alteration of the chemical composition of the atmosphere

- Construction of dams

- Deforestation and afforestation

- Removal of groundwater from wells

- Water abstraction from rivers

- Urbanization

- Effects on climate

Water Distribution on Earth

Earth is known as the "Blue Planet" because 71 percent of the Earth's surface is covered with water. Water also exists below land surface and as water vapor in the air. Water is a finite source. The bottled water that is consumed today might possibly be the same water that once trickled down the back of a wooly mammoth. The Earth is a closed system, meaning that very little matter, including water, ever leaves or enters the atmosphere; the water that was here billions of years ago is still here now. But, the Earth cleans and replenishes the water supply through the hydrologic cycle.

The earth has an abundance of water, but unfortunately, only a small percentage (about 0.3 percent), is even usable by humans. The other 99.7 percent is in the oceans, soils, icecaps, and floating in the atmosphere. Still, much of the 0.3 percent that is useable is unattainable. Most of the water used by humans comes from rivers. The visible bodies of water are referred to as surface water. The majority of fresh water is actually found underground as soil moisture and in aquifers. Groundwater can feed the streams, which is why a river can keep flowing even when there has been no precipitation. Humans can use both ground and surface water.

Distribution of the Water on Earth

- Ocean water: 97.2 percent

- Glaciers and other ice: 2.15 percent

- Groundwater: 0.61 percent

- Fresh water lakes: 0.009 percent

- Inland seas: 0.008 percent

- Soil Moisture: 0.005 percent

- Atmosphere: 0.001 percent

- Rivers: 0.0001 percent.

Surface Waters

Surface waters can be simply described as the water that is on the surface of the Earth. This includes the oceans, rivers and streams, lakes, and reservoirs. Surface waters are very important. They constitute approximately 80 percent of the water used on a daily basis. In 1990, the United States alone used approximately 327,000 billion gallons of surface water a day. Surface waters make up the majority of the water used for public supply and irrigation. It plays less of a role in mining and livestock industries. Oceans, which are the largest source of surface water, comprise approximately 97 percent of the Earth's surface water. However, since the oceans have high salinity, the water is not useful as drinking water. Efforts have been made to remove the salt from the water (desalination), but this is a very costly endeavor. Salt water is used in the mining process, in industry, and in power generation. The oceans also play a vital role in the hydrologic cycle, in regulating the global climate, and in providing habitats for thousands of marine species.

Rivers and streams constitute the flowing surface waters. The force of gravity naturally draws water from a higher altitude to a lower altitude. Rivers obtain their water from two sources: groundwater, and runoff. Rivers can obtain their water from the ground if they cut into the water table, the area in which the ground is saturated with water. This is known as base flow to the stream. Runoff flows downhill, first as small creeks, then gradually merging with other creeks and streams, increasing in size until a river has formed. These small creeks, or tributaries, where the river begins are known as the headwaters. Springs from confined aquifers also can contribute to rivers.

A river will eventually flow into an ocean. A river's length can be difficult to determine, especially if it has numerous tributaries. The USGS Web site defines a river's length as "the distance to the outflow point from the original headwaters where the name defines the complete length." In order for water to flow, there must be land upgradient of the river, that is land that is at a higher elevation than the river. The land that is upgradient of any point on the river is known as the drainage basin or watershed. Ridges of higher land, such as the Continental Divide, separate two drainage basins. Flowing water is extremely powerful and plays an important role in creating the landscape and in humans' lives. Flowing water is used for numerous reasons including irrigation and hydroelectric power production. Rivers erode the landscape and change the topography of the Earth by carving canyons and transporting soil and sediment to create fertile plains. Rivers carry soil and sediment that have been washed into the river when it rains or snow melts. The faster the water moves, the larger the particle size the river is capable of carrying. The USGS measures how much sediment a river carries by measuring the streamflow, or the amount of water flowing past a given site; and the sediment concentration. Sediment in the river can be helpful and harmful. Sediment, when deposited on the banks and in the flood plain, makes excellent farmlands. However, sediment can harm and even destroy dams, reservoirs and the life in the stream. Also, during floods, these sediments can be left behind as sticky, smelly mud in unwanted places.

Measuring the streamflow is accomplished by determining the stream stage and the stream discharge. The stream stage, or datum, is the height of the water surface, in feet, above an arbitrary reference point. The stream discharge is a measurement of the amount of water that is flowing at a particular point in time. It is measured in cubic feet per second. A discharge measurement determines the amount of water that is flowing in the river at any given stream stage. In order to make this measurement, the width of the river, and the water's depth and velocity at various points must be measured at several different stream stages. A cross-section of the river is divided into intervals and the area of each interval is calculated. If the velocity was measured at different depths on the same vertical interval, then the velocity is averaged. To determine the discharge for the interval, the area is multiplied by the velocity. To find the entire stream's discharge, an average of all the intervals' discharges is calculated. It is important to take discharge measurements of the stream at various stream stages, even flood stage.

A river reaches flood stage when the river overflows its banks. The flood stage can be determined by measuring the gage height, or simply the height of the water in the stream measured from the river's bottom. The streamflow can increase exponentially as the gage height increases. Thus, a small increase in gage height may indicate that a river has reached its flood stage. Floods are a fairly common, yet dangerous, natural disaster.

They normally occur because a storm or rapid snow melt has produced more runoff than a stream can carry. Dams failing, landslides blocking stream channels, and high tides are some other causes of flooding. Weather patterns can greatly influence when and where flooding will occur. By studying these patterns, geologists can determine the susceptibility of a region to having a flood at certain times of the year. The recurrence interval, measured in years, describes the magnitude of a flood. Changes in the drainage basin, such as harvesting timber or housing developments, can change the magnitude of a flood. The normally dry land that becomes covered with water during a flood is known as the flood plain. Restrictions on land use in flood plains is regulated by flood-plain zoning. Dams and levees have been built to help reduce damage caused by floods.

When flowing water travels to an area of land that is completely surrounded by higher land, a lake is formed. The water is not trapped in this low area, the water just escapes at a slower rate than the rate of incoming water. Lakes can vary greatly in area, depth, and water type. Most lakes are fresh water, however some, such as the Great Salt Lake and the Dead Sea, are salt water. Contrary to common belief, a reservoir is not the same as a lake. A reservoir is a manmade lake caused by a river being dammed. The water in a reservoir is very slow moving compared to the river. Therefore, the majority of the sediments that the river was carrying settle to the bottom of the reservoir. A reservoir will eventually fill up with sediment and mud and become unusable.

The Water Cycle

The hydrologic cycle or water cycle is a graphic representation of how water is recycled through the environment. Water molecules remain constant, though they may change between solid, liquid, and gas forms. Drops of water in the ocean evaporate, which is the process of liquid water becoming water vapor. Evaporation can occur from water surfaces, land surfaces, and snow fields into the air as water vapor. Moisture in the air can condensate, which is the process of water vapor in the air turning into liquid water. Water drops on the outside of a cold glass of water are condensed water. Condensation is the opposite process of evaporation. Water vapor condenses on tiny particles of dust, smoke, and salt crystals to become part of a cloud. After a while, the water droplets combine with other droplets and fall to Earth in the form of precipitation (rain, snow, hail, sleet, dew, and frost). Once the precipitation has fallen to Earth, it may go into an aquifer as groundwater or the drop may stay above ground as surface water. The hydrologic cycle is an important concept to understand. Water has so many uses on Earth, such as human and animal consumption, power production, and industrial and agricultural needs. Precipitation—in the form of rain and snow—also is an important thing to understand. It is the main way that the water in the skies comes down to Earth, where it fills the lakes and rivers, recharges the underground aquifers, and provides drinks to plants and animals. Different amounts of precipitation fall on different areas of the Earth at different rates and at various times of the year.

One problem facing the cycle of water on Earth is water contamination. Chemicals that go into the water often are very difficult, if not impossible, to remove. One potential source of contamination of water is runoff, the overland flow of water. While precipitation causes the runoff to occur, stripping vegetation from land can add to the runoff in a particular area. The sediment and soil from these areas, not to mention any pesticides or fertilizers that are present, are washed into the streams, oceans, and lakes. What happens to the rain after it falls depends on many factors, such as the intensity and duration of rainfall, the topography of the land, soil conditions, amount of urbanization, and density of vegetation. A common misconception about rain that it is tear-shaped, when in actuality it is shaped more like a hamburger bun. Rain drops also are different sizes, due to the initial difference in particle size and the different rate of coalescence.

Glaciers and Ice Caps

Glaciers and icecaps are referred to as storehouses for fresh water. They cover 10 percent of the world's land mass. These glaciers are primarily located in Greenland and Antarctica. The glaciers in Greenland almost cover the entire land mass. Glaciers begin forming because of snowfall accumulation. When snowfall exceeds the rate of melting in a certain area, glaciers begin to form. This melting occurs in the summer. The weight of snow accumulating compresses the snow to form ice. Because these glaciers are so heavy, they can slowly move their way down hills.

Glaciers affect the topography of the land in some areas. Ancient glaciers formed lakes and valleys. The Great Lakes are an example of this. Glaciers range in length from less than the size of a football field to hundreds of miles long. They also can reach up to 2 miles thick. Glaciers melting can have a tremendous effect on the sea level. If all of the glaciers were to melt today, the sea would rise an estimated 260 feet, according to the USGS. Glaciers have had a tremendous effect on the formation of the Earth's surface and are still influencing the topography everyday.

Groundwater

Groundwater is defined as water that is found beneath the surface of the Earth in conditions of 100 percent saturation (if it is less than 100 percent saturation, then the water is considered soil moisture). Ninety-eight percent of Earth's available fresh water is groundwater. It is about 60 times as plentiful as the fresh water found in lakes and streams. Water in the ground travels through pores in soil and rock, and in fractures and weathered areas of bedrock. The amount of pore space present in rock and soil is known as porosity. The ability to travel through the rock or soil is known as permeability. The permeability and porosity measurements in rock and/or soil can determine the amount of water that can flow through that particular medium. A "high" permeability and porosity value means that the water can travel quickly.

Groundwater can be found in aquifers. An aquifer is a body of water-saturated sediment or rock in which water can move readily. There are two main types of aquifers: unconfined and confined. An unconfined aquifer is a partially or fully filled aquifer that is exposed to the surface of the land. Because this aquifer is in contact with the atmosphere, it is impacted by meteoric water and any kind of surface contamination. There is not an impermeable layer to protect this aquifer. In contrast, a confined aquifer is an aquifer that has a confining layer that separates it from the land surface. This aquifer is filled with pressurized water (due to the confining layer). If the water is pressurized at a high enough value, when a well is drilled into the confining aquifer, water rises above the surface of the ground. This is known as a flowing water well. The pressure of the water is called the hydraulic head. Groundwater movement, or velocity, is measured in feet (or meters) per second.

In some areas, the bedrock has low permeability and porosity levels, yet groundwater can still travel in the aquifers. Groundwater can travel through fractures in the rock or through areas that are weathered. Limestone, for example, weathers in solution, creating underground cavities and cavern systems. At the land surface, these areas are known as "karst". The voids in the rock, created as limestone goes into solution, can cause collapses at the land surface. These collapses are known as sinkholes. Sinkholes are often a direct conduit to the groundwater and areas where contamination can easily infiltrate the aquifers. Sinkhole areas also can have land subsidence as mass wasting occurs in areas with a sudden change in slope and contact with water. Land subsidence may or may not be noticeable in some areas because it appears as hills and valleys (due to the very large size). As groundwater becomes more of a source for drinking water, the problem of sinkholes and land subsidence could increase.

Porosity and permeability of the sediment, soil, and bedrock in the area also affects the recharge rate of the groundwater. This means that in some areas, the groundwater can be pumped out faster than it can replenish itself. This creates a number of problems. One of these problems is called "drawdown," a lowering of the aquifer near a pumping well. This can occur in areas where the well is pumping faster than the groundwater aquifer is recharged. Drawdown creates voids in the bedrock and can lead to additional land subsidence or sinkholes (as there is no longer water present and the void cannot hold the weight of the material above and collapses).

Physical Oceanography

Physical oceanography is the study of the physics of marine systems. It includes the distribution of temperature and salinity, water mass formation and movement, ocean currents, interior and surface mixing, energy inputs and dissipation, surface and internal waves, and surface and internal tides.

This encompasses a very broad range of processes that can be characterized by the time and space scales over which they vary. On the rapidly varying end of the scale, there are turbulent eddies with durations of seconds and spatial scales of centimeters. At somewhat longer scales, there are propagating surface and internal gravity waves with periods of seconds to hours and wavelengths of meters to kilometers. Astronomical forces generate tides which propagate on the rotating earth as waves with periods predominantly near 12 and 24 hours and wavelengths of thousands of kilometers. At intermediate scales, there are horizontal eddies, fronts and coastal currents that vary on time scales of days to months and spatial scales of one to hundreds of kilometers. At the slowly varying end of the scale there are wind-forced and thermodynamically driven ocean currents with time scales of days to centuries and spatial scales of tens to thousands of kilometers. Transfers of momentum, heat and salt occur within the ocean and across the air-sea interface on all of these space and time scales. One of the intriguing aspects of physical oceanography is the overlap and interaction between the various physical processes. For example, processes that occur on very short and intermediate scales determine the water motion, temperature, salinity and other properties on very large scales. Large-scale water properties in the ocean are mixed by turbulent eddies that occur on vertical scales of centimeters and seconds and horizontal scales of kilometers and days. Vertical turbulent mixing can be enhanced by the cascade of energy from internal gravity waves that have vertical and horizontal scales of tens of meters and hours. A primary mechanism for the generation of internal waves is the interaction of ocean tides with bottom topography. It is thus apparent that a comprehensive understanding of the large-scale ocean circulation requires consideration of the full range of physical processes occurring in the ocean.

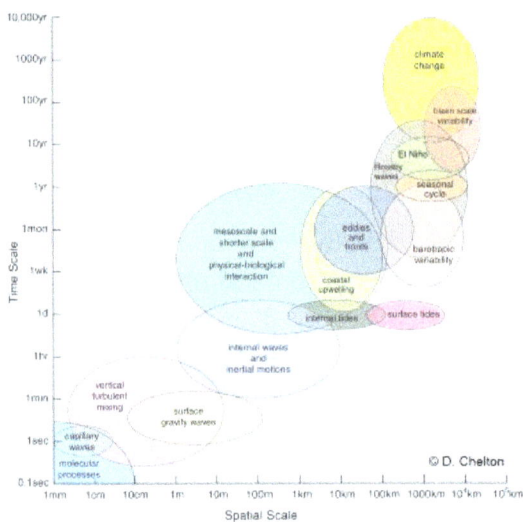

Physical Setting

Roughly 97% of the planet's water is in its oceans, and the oceans are the source of the vast majority of water vapor that condenses in the atmosphere and falls as rain or snow on the continents. The tremendous heat capacity of the oceans moderates the planet's climate, and its absorption of various gases affects the composition of the atmosphere. The ocean's influence extends even to the composition of volcanic rocks through seafloor metamorphism, as well as to that of volcanic gases and magmas created at subduction zones.

The oceans are far deeper than the continents are tall; examination of the Earth's hypsographic curve shows that the average elevation of Earth's landmasses is only 840 metres (2,760 ft), while the ocean's average depth is 3,800 metres (12,500 ft). Though this apparent discrepancy is great, for both land and sea, the respective extremes such as mountains and trenches are rare.

Area, volume plus mean and maximum depths of oceans (excluding adjacent seas)

Body	Area (10⁶km²)	Volume (10⁶km³)	Mean depth (m)	Maximum (m)
Pacific Ocean	165.2	707.6	4282	-11033
Atlantic Ocean	82.4	323.6	3926	-8605
Indian Ocean	73.4	291.0	3963	-8047
Southern Ocean	20.3			-7235
Arctic Ocean	14.1		1038	
Caribbean Sea	2.8			-7686

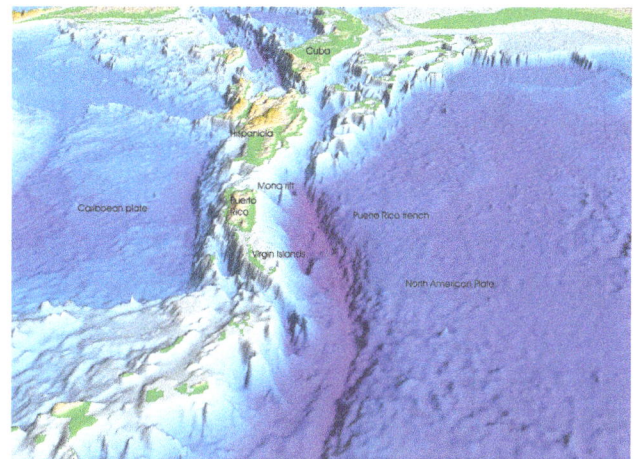

Perspective view of the sea floor of the Atlantic Ocean and the
Caribbean Sea. The purple sea floor at the center of
the view is the Puerto Rico Trench

Temperature, Salinity and Density

Because the vast majority of the world ocean's volume is deep water, the mean temperature of seawater is low; roughly 75% of the ocean's volume has a temperature from 0° – 5 °C (Pinet 1996). The same percentage falls in a salinity range between 34–35 ppt (3.4–3.5%) (Pinet 1996). There is still quite a bit of variation, however. Surface temperatures can range from below freezing near the poles to 35 °C in restricted tropical seas, while salinity can vary from 10 to 41 ppt (1.0–4.1%).

The vertical structure of the temperature can be divided into three basic layers, a surface mixed layer, where gradients are low, a thermocline where gradients are high, and a poorly stratified abyss.

In terms of temperature, the ocean's layers are highly latitude-dependent; the thermocline is pronounced in the tropics, but nonexistent in polar waters (Marshak 2001). The halocline usually lies near the surface, where evaporation raises salinity in the tropics, or meltwater dilutes it in polar regions. These variations of salinity and temperature with depth change the density of the seawater, creating the pycnocline.

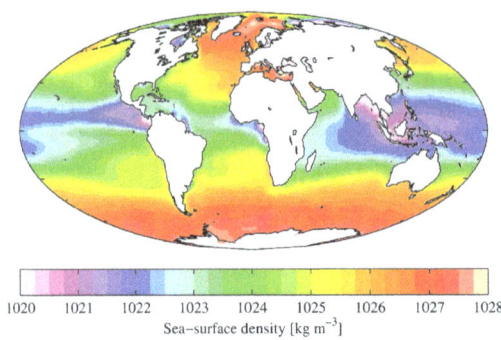

WOA surface density

Circulation

Energy for the ocean circulation (and for the atmospheric circulation) comes from solar radiation and gravitational energy from the sun and moon. The amount of sunlight absorbed at the surface varies strongly with latitude, being greater at the equator than at the poles, and this engenders fluid motion in both the atmosphere and ocean that acts to redistribute heat from the equator towards the poles, thereby reducing the temperature gradients that would exist in the absence of fluid motion. Perhaps three quarters of this heat is carried in the atmosphere; the rest is carried in the ocean.

The atmosphere is heated from below, which leads to convection, the largest expression of which is the Hadley circulation. By contrast the ocean is heated from above, which tends to suppress convection. Instead ocean deep water is formed in polar regions where cold salty waters sink in fairly restricted areas. This is the beginning of the thermohaline circulation.

Oceanic currents are largely driven by the surface wind stress; hence the large-scale atmospheric circulation is important to understanding the ocean circulation. The Hadley circulation leads to Easterly winds in the tropics and Westerlies in mid-latitudes. This leads to slow equatorward flow throughout most of a subtropical ocean basin (the Sverdrup balance). The return flow occurs in an intense, narrow, poleward western boundary current. Like the atmosphere, the ocean is far wider than it is deep, and hence horizontal motion is in general much faster than vertical motion. In the southern hemisphere there is a continuous belt of ocean, and hence the mid-latitude westerlies force the strong Antarctic Circumpolar Current. In the northern hemisphere the land masses prevent this and the ocean circulation is broken into smaller gyres in the Atlantic and Pacific basins.

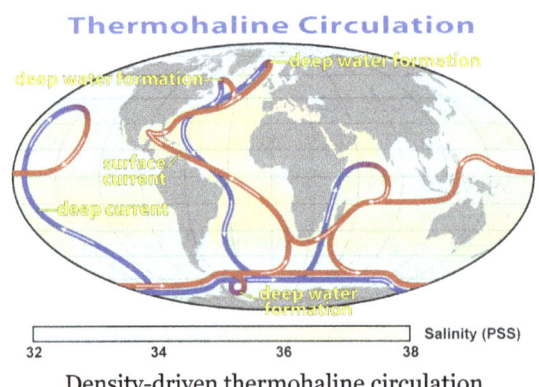

Density-driven thermohaline circulation

Coriolis Effect

The Coriolis effect results in a deflection of fluid flows (to the right in the Northern Hemisphere and left in the Southern Hemisphere). This has profound effects on the flow of the oceans. In particular it means the flow goes *around* high and low pressure systems, permitting them to persist for long periods of time. As a result, tiny variations in pressure can produce measurable currents. A slope of one part in one million in sea surface height, for example, will result in a current of 10 cm/s at mid-latitudes. The fact that the Coriolis effect is largest at the poles and weak at the equator results in sharp, relatively steady western boundary currents which are absent on eastern boundaries.

Ekman Transport

Ekman transport results in the net transport of surface water 90 degrees to the right of the wind in the Northern Hemisphere, and 90 degrees to the left of the wind in the Southern Hemisphere. As the wind blows across the surface of the ocean, it "grabs" onto a thin layer of the surface water. In turn, that thin sheet of water transfers motion energy to the thin layer of water under it, and so on. However, because of the Coriolis Effect, the direction of travel of the layers of water slowly move farther and farther to the right as they get deeper in the Northern Hemisphere, and to the left in the Southern Hemisphere. In most cases, the very bottom layer of water affected by the wind is at a depth of 100 m – 150 m and is traveling about 180 degrees, completely opposite of the direction that the wind is blowing. Overall, the net transport of water would be 90 degrees from the original direction of the wind.

Langmuir Circulation

Langmuir circulation results in the occurrence of thin, visible stripes, called windrows on the surface of the ocean parallel to the direction that the wind is blowing. If the wind is blowing with more than 3 m s^{-1}, it can create parallel windrows alternating upwelling and downwelling about 5–300 m apart. These windrows are created by adjacent ovular water cells (extending to about 6 m (20 ft) deep) alternating rotating clockwise and counterclockwise. In the convergence zones debris, foam and seaweed accumulates, while at the divergence zones plankton are caught and carried to the surface. If there are many plankton in the divergence zone fish are often attracted to feed on them.

Hurricane Isabel east of the Bahamas on
15 September 2003

Ocean–atmosphere Interface

At the ocean-atmosphere interface, the ocean and atmosphere exchange fluxes of heat, moisture and momentum.

Heat

The important heat terms at the surface are the sensible heat flux, the latent heat flux, the incoming solar radiation and the balance of long-wave (infrared) radiation. In general, the tropical oceans will tend to show a net gain of heat, and the polar oceans a net loss, the result of a net transfer of energy polewards in the oceans.

The oceans' large heat capacity moderates the climate of areas adjacent to the oceans, leading to a maritime climate at such locations. This can be a result of heat storage in summer and release in winter; or of transport of heat from warmer locations: a particularly notable example of this is Western Europe, which is heated at least in part by the north atlantic drift.

Momentum

Surface winds tend to be of order meters per second; ocean currents of order centimeters per second. Hence from the point of view of the atmosphere, the ocean can be considered effectively stationary; from the point of view of the ocean, the atmosphere imposes a significant wind stress on its surface, and this forces large-scale currents in the ocean.

Through the wind stress, the wind generates ocean surface waves; the longer waves have a phase velocity tending towards the wind speed. Momentum of the surface winds is transferred into the energy flux by the ocean surface waves. The increased roughness of the ocean surface, by the presence of the waves, changes the wind near the surface.

Moisture

The ocean can gain moisture from rainfall, or lose it through evaporation. Evaporative loss leaves the ocean saltier; the Mediterranean and Persian Gulf for example have strong evaporative loss; the resulting plume of dense salty water may be traced through the Straits of Gibraltar into the Atlantic Ocean. At one time, it was believed that evaporation/precipitation was a major driver of ocean currents; it is now known to be only a very minor factor.

Planetary Waves

Kelvin Waves

A Kelvin wave is any progressive wave that is channeled between two boundaries or opposing forces (usually between the Coriolis force and a coastline or the equator). There are two types, coastal and equatorial. Kelvin waves are gravity driven and non-dispersive. This means that Kelvin waves can retain their shape and direction over long periods of time. They are usually created by a sudden shift in the wind, such as the change of the trade winds at the beginning of the El Niño-Southern Oscillation.

Coastal Kelvin waves follow shorelines and will always propagate in a counterclockwise direction in the Northern hemisphere (with the shoreline to the right of the direction of travel) and clockwise in the Southern hemisphere.

Equatorial Kelvin waves propagate to the east in the Northern and Southern hemispheres, using the equator as a guide.

Kelvin waves are known to have very high speeds, typically around 2–3 meters per second. They have wavelengths of thousands of kilometers and amplitudes in the tens of meters.

Rossby Waves

Rossby waves, or planetary waves are huge, slow waves generated in the troposphere by temperature differences between the ocean and the continents. Their major restoring force is the change in Coriolis force with latitude. Their wave amplitudes are usually in the tens of meters and very large wavelengths. They are usually found at low or mid latitudes.

There are two types of Rossby waves, barotropic and baroclinic. Barotropic Rossby waves have the highest speeds and do not vary vertically. Baroclinic Rossby waves are much slower.

The special identifying feature of Rossby waves is that the phase velocity of each individual wave always has a westward component, but the group velocity can be in any direction. Usually the shorter Rossby waves have an eastward group velocity and the longer ones have a westward group velocity.

Climate Variability

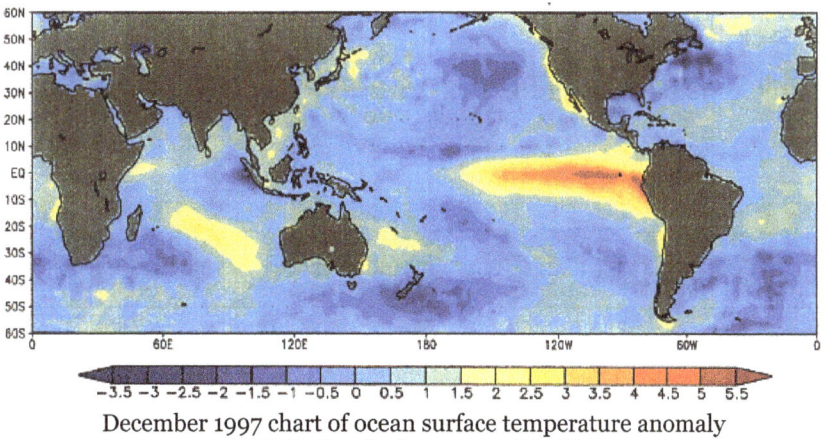

December 1997 chart of ocean surface temperature anomaly
[°C] during the last strong El Niño

The interaction of ocean circulation, which serves as a type of heat pump, and biological effects such as the concentration of carbon dioxide can result in global climate changes on a time scale of decades. Known climate oscillations resulting from these interactions, include the Pacific decadal oscillation, North Atlantic oscillation, and Arctic oscillation. The oceanic process of thermohaline circulation is a significant component of heat redistribution across the globe, and changes in this circulation can have major impacts upon the climate.

La Niña–El Niño

Antarctic Circumpolar Wave

This is a coupled ocean/atmosphere wave that circles the Southern Ocean about every eight years. Since it is a wave-2 phenomenon (there are two peaks and two troughs in a latitude circle) at each fixed point in space a signal with a period of four years is seen. The wave moves eastward in the direction of the Antarctic Circumpolar Current.

Ocean Currents

Among the most important ocean currents are the:

- Antarctic Circumpolar Current
- Deep ocean (density-driven)
- Western boundary currents
  - o   Gulf Stream
  - o   Kuroshio Current
  - o   Labrador Current
  - o   Oyashio Current
  - o   Agulhas Current
  - o   Brazil Current
  - o   East Australia Current
- Eastern Boundary currents California Current
- Canary Current
- Peru Current
- Benguela Current

Antarctic Circumpolar

The ocean body surrounding the Antarctic is currently the only continuous body of water where there is a wide latitude band of open water. It interconnects the Atlantic, Pacific and Indian oceans, and provide an uninterrupted stretch for the prevailing westerly winds to significantly increase wave amplitudes. It is generally accepted that these prevailing winds are primarily responsible for the circumpolar current transport. This current is now thought to vary with time, possibly in an oscillatory manner.

Deep Ocean

In the Norwegian Sea evaporative cooling is predominant, and the sinking water mass, the North Atlantic Deep Water (NADW), fills the basin and spills southwards through crevasses in the submarine sills that connect Greenland, Iceland and Britain. It then flows along the western boundary of the Atlantic with some part of the flow moving eastward along the equator and then poleward into the ocean basins. The NADW is entrained into the Circumpolar Current, and can be traced into the Indian and Pacific basins. Flow from the Arctic Ocean Basin into the Pacific, however, is blocked by the narrow shallows of the Bering Strait.

Western Boundary

An idealised subtropical ocean basin forced by winds circling around a high pressure (anticyclonic) systems such as the Azores-Bermuda high develops a gyre circulation with slow steady flows towards the equator in the interior. As discussed by Henry Stommel, these flows are balanced in the region of the western boundary, where a thin fast polewards flow called a western boundary current develops. Flow in the real ocean is more complex, but the Gulf stream, Agulhas and Kuroshio are examples of such currents. They are narrow (approximately 100 km across) and fast (approximately 1.5 m/s).

Equatorwards western boundary currents occur in tropical and polar locations, e.g. the East Greenland and Labrador currents, in the Atlantic and the Oyashio. They are forced by winds circulation around low pressure (cyclonic).

Gulf Stream

The Gulf Stream, together with its northern extension, North Atlantic Current, is a powerful, warm, and swift Atlantic Ocean current that originates in the Gulf of Mexico, exits through the Strait of Florida, and follows the eastern coastlines of the United States and New found land to the northeast before crossing the Atlantic Ocean.

Kuroshio

The Kuroshio Current is an ocean current found in the western Pacific Ocean off the east coast of Taiwan and flowing northeastward past Japan, where it merges with the easterly drift of the North Pacific Current. It is analogous to the Gulf Stream in the Atlantic Ocean, transporting warm, tropical water northward towards the polar region.

Heat Flux

Heat Storage

Ocean heat flux is a turbulent and complex system which utilizes atmospheric measurement techniques such as eddy covariance to measure the rate of heat transfer expressed in the unit of joules or watts per second. Heat flux is the difference in temperature between two points through which the heat passes. Most of the Earth's heat storage is within its seas with smaller fractions of the heat transfer in processes such as evaporation, radiation, diffusion, or absorption into the sea floor. The majority of the ocean heat flux is through advection or the movement of the ocean's currents. For example, the majority of the warm water movement in the south Atlantic is thought to have originated in the Indian Ocean. Another example of advection is the nonequatorial Pacific heating which results from subsurface processes related to atmospheric anticlines. Recent warming observations of Antarctic Bottom Water in the Southern Ocean is of concern to ocean scientists be-cause bottom water changes will effect currents, nutrients, and biota elsewhere. The international awareness of global warming has focused scientific research on this topic since the 1988 creation of the Intergovernmental Panel on Climate Change. Improved ocean observation, instrumentation, theory, and funding has increased scientific reporting on regional and global issues related to heat.

Sea Level Change

Tide gauges and satellite altimetry suggest an increase in sea level of 1.5–3 mm/yr over the past 100 years.

The IPCC predicts that by 2081-2100, global warming will lead to a sea level rise of 260 to 820 mm.

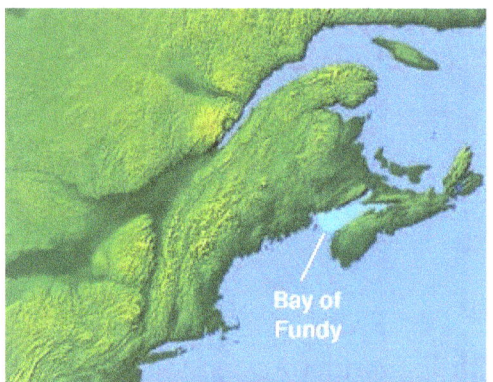

The Bay of Fundy is a bay located on the Atlantic coast of North America, on the northeast end of the Gulf of Maine between the provinces of New Brunswick and Nova Scotia

Rapid Variations

Tides

The rise and fall of the oceans due to tidal effects is a key influence upon the coastal areas. Ocean tides on the planet Earth are created by the gravitational effects of the Sun and Moon. The tides produced by these two bodies are roughly comparable in magnitude, but the orbital motion of the Moon results in tidal patterns that vary over the course of a month.

The ebb and flow of the tides produce a cyclical current along the coast, and the strength of this current can be quite dramatic along narrow estuaries. Incoming tides can also produce a tidal bore along a river or narrow bay as the water flow against the current results in a wave on the surface.

Tide and Current (Wyban 1992) clearly illustrates the impact of these natural cycles on the life-style and livelihood of Native Hawaiians tending coastal fishponds. *Aia ke ola ka hana* meaning . . . *Life is in labor.*

Tidal resonance occurs in the Bay of Fundy since the time it takes for a large wave to travel from the mouth of the bay to the opposite end, then reflect and travel back to the mouth of the bay coin-cides with the tidal rhythm producing the world's highest tides.

As the surface tide oscillates over topography, such as submerged seamounts or ridges, it gener-ates internal waves at the tidal frequency, which are known as internal tides.

Tsunamis

A series of surface waves can be generated due to large-scale displacement of the ocean water. These can be caused by sub-marine landslides, seafloor deformations due to earthquakes, or the impact of a large meteorite.

The waves can travel with a velocity of up to several hundred km/hour across the ocean surface, but in mid-ocean they are barely detectable with wavelengths spanning hundreds of kilometers.

Tsunamis, originally called tidal waves, were renamed because they are not related to the tides. They are regarded as shallow-water waves, or waves in water with a depth less than 1/20 their wavelength. Tsunamis have very large periods, high speeds, and great wave heights.

The primary impact of these waves is along the coastal shoreline, as large amounts of ocean water are cyclically propelled inland and then drawn out to sea. This can result in significant modifications to the coastline regions where the waves strike with sufficient energy.

The tsunami that occurred in Lituya Bay, Alaska on July 9, 1958 was 520 m (1,710 ft) high and is the biggest tsunami ever measured, almost 90 m (300 ft) taller than the Sears Tower in Chicago and about 110 m (360 ft) taller than the former World Trade Center in New York.

Surface Waves

The wind generates ocean surface waves, which have a large impact on offshore structures, ships, coastal erosion and sedimentation, as well as harbours. After their generation by the wind, ocean surface waves can travel (as swell) over long distances.

Ocean Current

Ocean current is stream made up of horizontal and vertical components of the circulation system of ocean waters that is produced by gravity, wind friction, and water density variation in different parts of the ocean. Ocean currents are similar to winds in the atmosphere in that they transfer significant amounts of heat from Earth's equatorial areas to the poles and thus play important roles in determining the climates of coastal regions. In addition, ocean currents and atmospheric circulation influence one another.

The general circulation of the oceans defines the average movement of seawater, which, like the atmosphere, follows a specific pattern. Superimposed on this pattern are oscillations of tides and waves, which are not considered part of the general circulation. There also are meanders and eddies that represent temporal variations of the general circulation. The ocean circulation pattern exchanges water of varying characteristics, such as temperature and salinity, within the interconnected network of oceans and is an important part of the heat and freshwater fluxes of the global climate. Horizontal movements are called currents, which range in magnitude from a few centimetres per second to as much as 4 metres (about 13 feet) per second. A characteristic surface speed is about 5 to 50 cm (about 2 to 20 inches) per second. Currents generally diminish in intensity with increasing depth. Vertical movements, often referred to as upwelling and downwelling, exhibit much lower speeds, amounting to only a few metres per month. As seawater is nearly incompressible, vertical movements are associated with regions of convergence and divergence in the horizontal flow patterns.

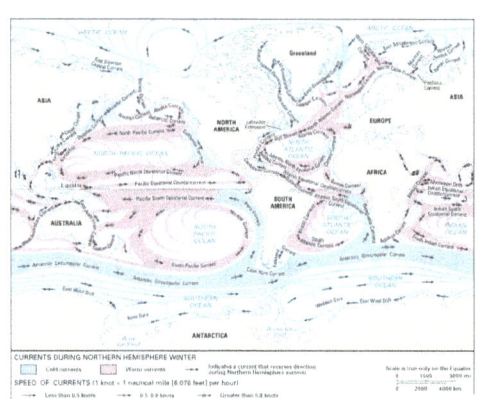 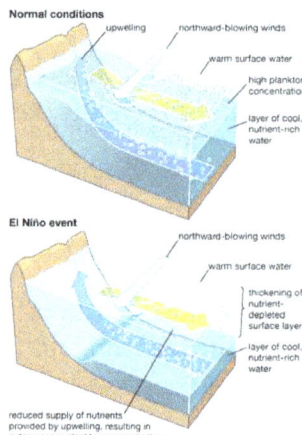

The upwelling process in the ocean along the coast of Peru. A thermocline and a nutricline separate the warm, nutrient-deficient upper layer from the cool, enriched layer below.

Under normal conditions (top), these interfaces are shallow enough that coastal winds can induce upwelling of the lower-layer nutrients to the surface, where they support an abundant ecosystem. During an El Niño event (bottom), the upper layer thickens so that the upwelled water contains fewer nutrients, thus contributing to a collapse of marine productivity.

Distribution Of Ocean Currents

Maps of the general circulation at the sea surface were originally constructed from a vast amount of data obtained from inspecting the residual drift of ships after course direction and speed are accounted for in a process called dead reckoning. This information is collected by satellite-tracked surface drifters at sea at present. The pattern is nearly entirely that of wind-driven circulation.

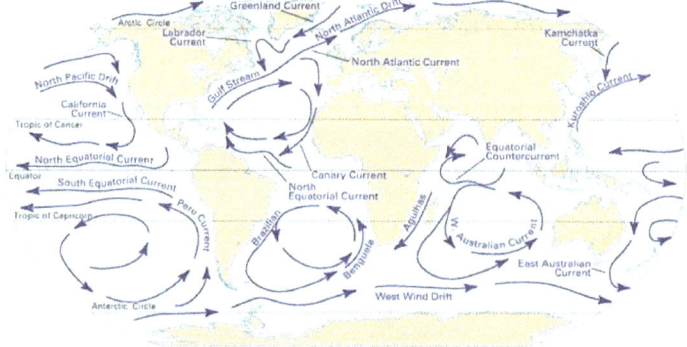

Major surface currents of the world's oceans. Subsurface currents also move vast amounts of water, but they are not known in such detail.

At the surface, aspects of wind-driven circulation cause the gyres (large anticyclonic current cells that spiral about a central point) to displace their centres westward, forming strong western boundary currents against the eastern coasts of the continents, such as the Gulf Stream–North Atlantic–Norway Current in the Atlantic Ocean and the Kuroshio–North Pacific Current in the Pacific

Ocean. In the Southern Hemisphere the counterclockwise circulation of the gyres creates strong eastern boundary currents against the western coasts of continents, such as the Peru (Humboldt) Current off South America, the Benguela Current off western Africa, and the Western Australia Current. The Southern Hemisphere currents are also influenced by the powerful, eastward-flowing, circumpolar Antarctic Current. It is a very deep, cold, and relatively slow current, but it carries a vast mass of water, about twice the volume of the Gulf Stream. The Peru and Benguela currents draw water from this Antarctic current and, hence, are cold. The Northern Hemisphere lacks continuous open water bordering the Arctic and so has no corresponding powerful circumpolar current, but there are small cold currents flowing south through the Bering Strait to form the Oya and Anadyr currents off eastern Russia and the California Current off western North America; others flow south around Greenland to form the cold Labrador and East Greenland currents. The Kuroshio–North Pacific and Gulf Stream–North Atlantic–Norway currents move warmer water into the Arctic Ocean via the Bering, Cape, and West Spitsbergen currents.

In the tropics the great clockwise and counter clockwise gyres flow westward as the Pacific North and South Equatorial currents, Atlantic North and South Equatorial currents, and the Indian South Equatorial Current. Because of the alternating monsoon climate of the northern Indian Ocean, the current in the northern Indian Ocean and the Arabian Sea alternates. Between these massive currents are narrow eastward-flowing counte currents.

Other smaller current systems found in certain enclosed seas or ocean areas are less affected by wind-driven circulation and more influenced by the direction of water inflow. Such currents are found in the Tasmanian Sea, where the southward-flowing East Australian Current generates counterclockwise circulation, in the northwestern Pacific, where the eastward-flowing Kuroshio–North Pacific current causes counterclockwise circulation in the Alaska Current and Aleutian Current (or Subarctic Current), in the Bay of Bengal, and in the Arabian Sea.

Deep-ocean circulation consists mainly of thermohaline circulation. The currents are inferred from the distribution of seawater properties, which trace the spreading of specific water masses. The distribution of density is also used to estimate the deep currents. Direct observations of subsurface currents are made by deploying current meters from bottom-anchored moorings and by setting out neutral buoyant instruments whose drift at depth is tracked acoustically.

Causes of Ocean Currents

The general circulation is governed by the equation of motion, one of the fundamental laws of mechanics developed by English physicist and mathematician Sir Isaac Newton that was applied to a continuous volume of water. This equation states that the product of mass and current acceleration equals the vector sum of all forces that act on the mass. Besides gravity, the most important forces that cause and affect ocean currents are horizontal pressure-gradient forces, Coriolis forces, and frictional forces. Temporal and inertial terms are generally of secondary importance to the general flow, though they become important for transient features such as meanders and eddies.

Pressure Gradients

The hydrostatic pressure, p, at any depth below the sea surface is given by the equation $p = g\rho z$, where g is the acceleration of gravity, ρ is the density of seawater, which increases with depth, and z is the depth below the sea surface. This is called the hydrostatic equation, which is a good approximation for the equation of motion for forces acting along the vertical. Horizontal differences in density (due to variations of temperature and salinity) measured along a specific depth cause the hydrostatic pressure to vary along a horizontal plane or geopotential surface, a surface perpendicular to the direction of the gravity acceleration. Horizontal gradients of pressure, though much smaller than vertical changes in pressure, give rise to ocean currents.

In a homogeneous ocean, which would have a constant potential density, horizontal pressure differences are possible only if the sea surface is tilted. In this case, surfaces of equal pressure, called isobaric surfaces, are tilted in the deeper layers by the same amount as the sea surface. This is referred to as the barotropic field of mass. The unchanged pressure gradient gives rise to a current speed independent of depth. The oceans of the world, however, are not homogeneous. Horizontal variations in temperature and salinity cause the horizontal pressure gradient to vary with depth. This is the baroclinic field of mass, which leads to currents that vary with depth. The horizontal pressure gradient in the ocean is a combination of these two mass fields.

The tilt, or topographic relief, of the isobaric surface marking sea surface (defined as p = 0) can be constructed from a three-dimensional density distribution using the hydrostatic equation. Since the absolute value of pressure is not measured at all depths in the ocean, the sea surface slope is presented relative to that of a deep isobaric surface; it is assumed that the deep isobaric surface is level. Since the wind-driven circulation attenuates with increasing depth, an associated decrease of isobaric tilt with increasing depth is expected. Representation of the sea surface relief relative to a deep reference surface is a good representation of the absolute shape of the sea surface. The total relief of the sea surface amounts to about 2 metres (about 6.5 feet), with "hills" in the subtropics and "valleys" in the polar regions. This pressure head drives the surface circulation.

Coriolis Effect

Earth's rotation about its axis causes moving particles to behave in a way that can only be understood by adding a rotational dependent force. To an observer in space, a moving body would continue to move in a straight line unless the motion were acted upon by some other force. To an Earth-bound observer, however, this motion cannot be along a straight line because the reference frame is the rotating Earth. This is similar to the effect that would be experienced by an observer standing on a large turntable if an object moved over the turntable in a straight line relative to the "outside" world. An apparent deflection of the path of the moving object would be seen. If the turntable rotated counterclockwise, the apparent deflection would be to the right of the direction of the moving object, relative to the observer fixed on the turntable.

This remarkable effect is evident in the behaviour of ocean currents. It is called the Coriolis force, named after Gustave-Gaspard Coriolis, a 19th-century French engineer and mathematician. For Earth, horizontal deflections due to the rotational induced Coriolis force act on particles moving in any horizontal direction. There also are apparent vertical forces, but these are of minor importance to ocean currents. Because Earth rotates from west to east about its axis, an observer in the Northern Hemisphere would notice a deflection of a moving body toward the right. In the Southern Hemisphere, this deflection would be toward the left. As a result, ocean currents move clockwise (anticyclonically) in the Northern Hemisphere and counterclockwise (cyclonically) in the Southern Hemisphere; Coriolis force deflects them about 45° from the wind direction, and at the Equator there would be no apparent horizontal deflection.

It can be shown that the Coriolis force always acts perpendicular to motion. Its horizontal component, Cf, is proportional to the sine of the geographic latitude (θ, given as a positive value for the Northern Hemisphere and a negative value for the Southern Hemisphere) and the speed, c, of the moving body. It is given by Cf = c ($2\omega \sin \theta$), where $\omega = 7.29 \times 10^{-5}$ radian per second is the angular velocity of Earth's rotation.

Frictional Forces

Movement of water through the oceans is slowed by friction, with surrounding fluid moving at a different velocity. A faster-moving fluid layer tends to drag along a slower-moving layer, and a slower-moving layer will tend to reduce the speed of a faster-moving layer. This momentum transfer between the layers is referred to as frictional forces. The momentum transfer is a product of turbulence that moves kinetic energy to smaller scales until at the tens-of-microns scale (1 micron = 1/1,000 mm) it is dissipated as heat. The wind blowing over the sea surface transfers momentum to the water. This frictional force at the sea surface (i.e., the wind stress) produces the wind-driven circulation. Currents moving along the ocean floor and the sides of the ocean also are subject to the influence of boundary-layer friction. The motionless ocean floor removes momentum from the circulation of the ocean waters.

Geostrophic Currents

For most of the ocean volume away from the boundary layers, which have a characteristic thickness of 100 metres (about 330 feet), frictional forces are of minor importance, and the equation of motion for horizontal forces can be expressed as a simple balance of horizontal pressure gradient and Coriolis force. This is called geostrophic balance.

On a non rotating Earth, water would be accelerated by a horizontal pressure gradient and would flow from high to low pressure. On the rotating Earth, however, the Coriolis force deflects the motion, and the acceleration ceases only when the speed, U, of the current is just fast enough to produce a Coriolis force that can exactly balance the horizontal pressure-gradient force. This geostrophic balance is given as dp/dx = v2ω sin θ, and dp/dy = –u2 sin, where dp/dx and dp/dy are the horizontal pressure gradient along the x-axis and y-axis, respectively, and u and v are the horizontal components of the velocity U along the x-axis and y-axis, respectively. From this balance it follows that the current direction must be perpendicular to the pressure gradient because the Coriolis force always acts perpendicular to the motion. In the Northern Hemisphere this direction is such that the high pressure is to the right when looking in current direction, while in the Southern Hemisphere it is to the left. This type of current is called a geostrophic current. The simple equation given above provides the basis for an indirect method of computing ocean currents. The relief of the sea surface also defines the streamlines (paths) of the geostrophic current at the surface relative to the deep reference level. The hills represent high pressure, and the valleys stand for low pressure. Clockwise rotation in the Northern Hemisphere with higher pressure in the centre of rotation is called anticyclonic motion. Counterclockwise rotation with lower pressure in its centre is cyclonic motion. In the Southern Hemisphere the sense of rotation is the opposite, because the effect of the Coriolis force has changed its sign of deflection.

Ekman Layer

The wind exerts stress on the ocean surface proportional to the square of the wind speed and in the direction of the wind, setting the surface water in motion. This motion extends to a depth of about 100 metres in what is called the Ekman layer, after the Swedish oceanographer V. Walfrid Ekman, who in 1902 deduced these results in a theoretical model constructed to help explain observations of wind drift in the Arctic. Within the oceanic Ekman layer the wind stress is balanced by the Coriolis force and frictional forces. The surface water is directed at an angle of 45° to the wind, to the right in the Northern Hemisphere and to the left in the Southern Hemisphere. With increasing depth in the boundary layer, the current speed is reduced, and the direction rotates farther away from the wind direction following a spiral form, becoming antiparallel to the surface flow at the base of the layer where the speed is 1/23 of the surface speed. This so-called Ekman spiral may be the exception rather than the rule, as the specific conditions are not often met, though deflection of a wind-driven surface current at somewhat smaller than 45° is observed when the wind field blows with a steady force and direction for the better part of a day. The average water particle within the Ekman layer moves at an angle of 90° to the wind; this movement is to the right of the wind direction in the Northern Hemisphere and to its left in the Southern Hemisphere. This phenomenon is called Ekman transport, and its effects are widely observed in the oceans.

Since the wind varies from place to place, so does the Ekman transport, forming convergence and divergence zones of surface water. A region of convergence forces surface water downward in a process called downwelling, while a region of divergence draws water from below into the surface Ekman layer in a process known as upwelling. Upwelling and downwelling also occur where the wind blows parallel to a coastline. The principal upwelling regions of the world are along the eastern boundary of the subtropical ocean waters, as, for example, the coastal region of Peru and northwestern Africa. Upwelling in these regions cools the surface water and brings nutrient-rich subsurface water into the sunlit layer of the ocean, resulting in a biologically productive region. Upwelling and high productivity also are found along divergence zones at the Equator and around Antarctica. The primary downwelling regions are in the subtropical ocean waters—e.g., the Sargasso Sea in the North Atlantic. Such areas are devoid of nutrients and are poor in marine life.

The vertical movements of ocean waters into or out of the base of the Ekman layer amount to less than 1 metre (about 3.3 feet) per day, but they are important since they extend the wind-driven effects into deeper waters. Within an upwelling region, the water column below the Ekman layer is drawn upward. This process, with conservation of angular momentum on the rotating Earth, induces the water column to drift toward the poles. Conversely, downwelling forces water into the water column below the Ekman layer, inducing drift toward the Equator. An additional consequence of upwelling and downwelling for stratified waters is to create a baroclinic field of mass. Surface water is less dense than deeper water. Ekman convergences have the effect of accumulating less dense surface water. This water floats above the surrounding water, forming a hill in sea level and driving an anticyclonic geostrophic current that extends well below the Ekman layer. Divergences do the opposite: they remove the less dense surface water, replacing it with denser, deeper water. This induces a depression in sea level with a cyclonic geostrophic current.

The ocean current pattern produced by the wind-induced Ekman transport is called the Sverdrup transport, after the Norwegian oceanographer H.U. Sverdrup, who formulated the basic theory in 1947. Several years later (1950) the American geophysicist and oceanographer Walter H. Munk and others expanded Sverdrup's work, explaining many of the major features of the wind-driven general circulation by using the mean climatological wind stress distribution at the sea surface as a driving force.

Two Types Of Ocean Circulation

Ocean circulation derives its energy at the sea surface from two sources that define two circulation types: (1) wind-driven circulation forced by wind stress on the sea surface, inducing a momentum exchange, and (2) thermohaline circulation driven by the variations in water density imposed at the sea surface by exchange of ocean heat and water with the atmosphere, inducing a buoyancy exchange. These two circulation types are not fully independent, since the sea-air buoyancy and momentum exchange are dependent on wind speed. The wind-driven circulation is the more vigorous of the two and is configured as gyres that dominate an ocean region. The wind-driven circulation is strongest in the surface layer. The thermohaline circulation is more sluggish, with a typical speed of 1 cm (0.4 inch) per second, but this flow extends to the seafloor and forms circulation patterns that envelop the global ocean.

Wind-driven Circulation

Wind stress induces a circulation pattern that is similar for each ocean. In each case, the wind-driven circulation is divided into gyres that stretch across the entire ocean: subtropical gyres extend from the equatorial current system to the maximum westerlies in a wind field near 50° latitude, and subpolar gyres extend poleward of the maximum westerlies. The depth penetration of the wind-driven currents depends on the intensity of ocean stratification: in those regions of strong stratification, such as the tropics, the surface currents extend to a depth of less than 1,000 metres (about 3,300 feet), and within the low-stratification polar regions the wind-driven circulation reaches all the way to the seafloor.

Equatorial Currents

At the Equator the currents are for the most part directed toward the west, the North Equatorial Current in the Northern Hemisphere and the South Equatorial Current in the Southern Hemisphere. Near the thermal equator, where the warmest surface water is found, there occurs the eastward-flowing Equatorial Counter Current. This current is slightly north of the geographic Equator, drawing the northern fringe of the South Equatorial Current to 5° N. The offset to the Northern Hemisphere matches a similar offset in the wind field. The east-to-west wind across the tropical ocean waters induces Ekman transport divergence at the Equator, which cools the surface water there.

At the geographic Equator a jetlike current is found just below the sea surface, flowing toward the east counter to the surface current. This is called the Equatorial Undercurrent. It attains speeds of more than 1 metre per second at a depth of nearly 100 metres. It is driven by higher sea level in the western margins of the tropical ocean, producing a pressure gradient, which in the absence of a horizontal Coriolis force drives a west-to-east current along the Equator. The wind field reverses the flow within the surface layer, inducing the South Equatorial Current.

Equatorial circulation undergoes variations following the irregular periods of roughly three to eight years of the Southern Oscillation (i.e., fluctuations of atmospheric pressure over the tropical

Indo-Pacific region). Weakening of the east-to-west wind during a phase of the Southern Oscillation allows warm water in the western margin to slip back to the east by increasing the flow of the Equatorial Counter Current. Surface water temperatures and sea level decrease in the west and increase in the east. This event is called El Niño. The combined El Niño/Southern Oscillation (ENSO) effect has received much attention because it is associated with global-scale climatic variability. In the tropical Indian Ocean the strong seasonal winds of the monsoons induce a similarly strong seasonal circulation pattern.

The Subtropical Gyres

The subtropical gyres are anticyclonic circulation features. The Ekman transport within these gyres forces surface water to sink, giving rise to the subtropical convergence near 20°–30° latitude. The centre of the subtropical gyre is shifted to the west. This westward intensification of ocean currents was explained by the American meteorologist and oceanographer Henry M. Stommel (1948) as resulting from the fact that the horizontal Coriolis force increases with latitude. This causes the poleward-flowing western boundary current to be a jetlike current that attains speeds of 2 to 4 metres (6.5 to 13 feet) per second. This current transports the excess heat of the low latitudes to higher latitudes. The flow within the equatorward-flowing interior and eastern boundary of the subtropical gyres is quite different. It is more of a slow drift of cooler water that rarely exceeds 10 cm (about 4 inches) per second. Associated with these currents is coastal upwelling that results from offshore Ekman transport.

The strongest of the western boundary currents is the Gulf Stream in the North Atlantic Ocean. It carries about 30 million cubic metres (1 billion cubic feet) of ocean water per second through the Straits of Florida and roughly 80 million cubic metres (2.8 billion cubic feet) per second as it flows past Cape Hatteras off the coast of North Carolina, U.S. Responding to the large-scale wind field over the North Atlantic, the Gulf Stream separates from the continental margin at Cape Hatteras. After separation it forms waves or meanders that eventually generate many eddies of warm and cold water. The warm eddies, composed of thermocline water normally found south of the Gulf Stream, are injected into the waters of the continental slope off the coast of the northeastern United States. They drift to the southwest at rates of approximately 5 to 8 cm (about 2 to 3 inches) per second, and after a year they rejoin the Gulf Stream north of Cape Hatteras. Cold eddies of slope water are injected into the region south of the Gulf Stream and drift to the southwest. After roughly two years they reenter the Gulf Stream just north of the Antilles islands. The path that they follow defines a clockwise-flowing recirculation gyre seaward of the Gulf Stream.

Among the other western boundary currents, the Kuroshio of the North Pacific is perhaps the most like the Gulf Stream, having a similar transport and array of eddies. The Brazil Current and the East Australian Current are relatively weak. The Agulhas Current has a transport close to that of the Gulf Stream. It remains in contact with the margin of Africa around the southern rim of the continent. It then separates from the margin and curls back to the Indian Ocean in what is called the Agulhas Retroflection. Not all the water carried by the Agulhas Current returns to the east; about 10 to 20 percent is injected into the South Atlantic Ocean as large eddies that slowly migrate across it.

The Subpolar Gyres

The subpolar gyres are cyclonic circulation features. The Ekman transport within these features forces upwelling and surface water divergence. In the North Atlantic the subpolar gyre consists of the North Atlantic Current at its equatorward side and the Norwegian Current that carries relatively warm water northward along the coast of Norway. The heat released from the Norwegian Current into the atmosphere maintains a moderate climate in northern Europe. Along the east coast of Greenland is the southward-flowing cold East Greenland Current. It loops around the southern tip of Greenland and continues flowing into the Labrador Sea. The southward flow that continues off the coast of Canada is called the Labrador Current. This current separates for the most part from the coast near Newfoundland to complete the subpolar gyre of the North Atlantic. Some of the cold water of the Labrador Current, however, extends farther south.

In the North Pacific the subpolar gyre is composed of the northward-flowing Alaska Current, the Aleutian Current, and the southward-flowing cold Oyashio Current. The North Pacific Current forms the separation between the subpolar and subtropical gyres of the North Pacific.

In the Southern Hemisphere the subpolar gyres are less defined. Large cyclonic flowing gyres lie poleward of the Antarctic Circumpolar Current and can be considered counterparts to the Northern Hemispheric subpolar gyres. The best-formed is the Weddell Gyre of the South Atlantic sector of the Southern Ocean. The Antarctic coastal current flows toward the west. The northward-flowing current off the east coast of the Antarctic Peninsula carries cold Antarctic coastal water into the circumpolar belt. Another cyclonic gyre occurs north of the Ross Sea.

The Antarctic Circumpolar Current

The Southern Ocean links the major oceans by a deep circumpolar belt in the 50°–60° S range. In this belt flows the Antarctic Circumpolar Current from west to east, encircling the globe at high latitudes. It transports 125 million cubic metres (4.4 billion cubic feet) of seawater per second over a path of about 24,000 km (about 14,900 miles) and is the most important factor in diminishing the differences between oceans. The Antarctic Circumpolar Current is not a well-defined single-axis current but rather consists of a series of individual currents separated by frontal zones. It reaches the seafloor and is guided along its course by the irregular bottom topography. Large meanders and eddies develop in the current as it flows. These features induce poleward transfer of heat, which may be significant in balancing the oceanic heat loss to the atmosphere above the Antarctic region farther south.

Thermohaline Circulation

The general circulation of the oceans consists primarily of the wind-driven currents. These, however, are superimposed on the much more sluggish circulation driven by horizontal differences in temperature and salinity—namely, the thermohaline circulation. The thermohaline circulation reaches down to the seafloor and is often referred to as the deep, or abyssal, ocean circulation. Measuring seawater temperature and salinity distribution is the chief method of studying the deep-flow patterns. Other properties also are examined; for example, the concentrations of oxygen, carbon-14, and such synthetically produced compounds as chlorofluorocarbons are measured to obtain resident times and spreading rates of deep water.

Introduction to Physical Geography

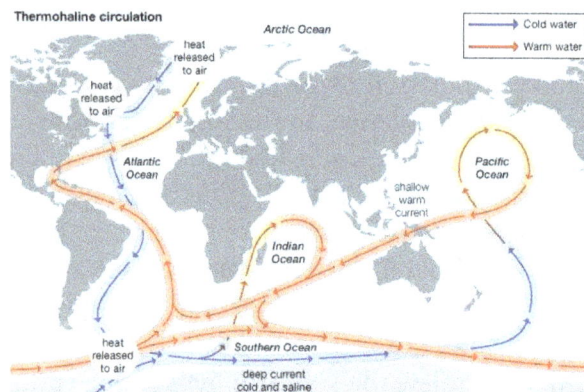

Thermohaline circulation transports and mixes the water of the oceans.

In the process it transports heat, which influences regional climate patterns. The density of seawater is determined by the temperature and salinity of a volume of seawater at a particular location. The difference in density between one location and another drives the thermohaline circulation.

In some areas of the ocean, generally during the winter season, cooling or net evaporation causes surface water to become dense enough to sink. Convection penetrates to a level where the density of the sinking water matches that of the surrounding water. It then spreads slowly into the rest of the ocean. Other water must replace the surface water that sinks. This sets up the thermohaline circulation. The basic thermohaline circulation is one of sinking of cold water in the polar regions, chiefly in the northern North Atlantic and near Antarctica. These dense water masses spread into the full extent of the ocean and gradually upwell to feed a slow return flow to the sinking regions. A theory for the thermohaline circulation pattern was proposed by Stommel and Arnold Arons in 1960.

In the Northern Hemisphere the primary region of deep water formation is the North Atlantic; minor amounts of deep water are formed in the Red Sea and Persian Gulf. A variety of water types contribute to the so-called North Atlantic Deep Water. Each one of them differs, though they share a common attribute of being relatively warm (greater than 2° C) and salty (greater than 34.9 parts per thousand) compared with the other major producer of deep and bottom water, the Southern Ocean (0° C and 34.7 parts per thousand). North Atlantic Deep Water is primarily formed in the Greenland and Norwegian seas, where cooling of the salty water introduced by the Norwegian Current induces sinking. This water spills over the rim of the ridge that stretches from Greenland to Scotland, extending to the seafloor to the south as a convective plume. It then flows southward, pressed against the western edge of the North Atlantic. Additional deep water is formed in the Labrador Sea. This water, somewhat less dense than the overflow water from the Greenland and Norwegian seas, has been observed sinking to a depth of 3,000 metres (about 9,800 feet) within convective features referred to as chimneys. Vertical velocities as high as 10 cm per second have been observed within these convective features. A third variety of North Atlantic Deep Water is derived from net evaporation within the Mediterranean Sea. This draws surface water into the Mediterranean through the Strait of Gibraltar. The mass of salty water formed within the Mediterranean exits as a deeper stream. It descends to depths of approximately 1,000 metres in the North Atlantic Ocean, forming the uppermost layer of North Atlantic Deep Water. The outflow in the Strait of Gibraltar reaches as high as 2 metres per second, but its total transport amounts to only 5 percent of the total North Atlantic Deep Water formed. The outflow of the Mediterranean plays a significant role in boosting the salinity of North Atlantic Deep Water.

The blend of North Atlantic Deep Water, with a total formation rate of 15 to 20 million cubic metres (530 to 706 million cubic feet) per second, quickly ventilates the Atlantic Ocean, resulting in a residence time of less than 200 years. The deep water spreads away from its source along the western side of the Atlantic Ocean and, on reaching the Antarctic Circumpolar Current, spreads into the Indian and Pacific oceans. The sinking of North Atlantic Deep Water is compensated for by the slow upwelling of deep water, mainly in the Southern Ocean, to replenish the upper stratum of water that has descended as North Atlantic Deep Water. North Atlantic Deep Water exported to the other oceans must be balanced by the inflow of upper-layer water into the Atlantic. Some water returns as cold, low-salinity Pacific water through the Drake Passage in the form of what is known as Antarctic Intermediate Water, and some returns as warm salty thermocline water from the Indian Ocean around the southern rim of Africa.

Remnants of North Atlantic Deep Water mix with Southern Ocean water to spread along the seafloor into the North Pacific Ocean. Here it upwells to a level of 2,000–3,000 metres (about 6,500–9,800 feet) and returns to the south lower in salinity and oxygen but higher in nutrient concentrations as North Pacific Deep Water. This North Pacific Deep Water is eventually swept eastward with the Antarctic Circumpolar Current. Modification of deep water in the North Pacific is the direct consequence of vertical mixing, which carries into the deep ocean the low salinity properties of North Pacific Intermediate Water. The latter is formed in the northwestern Pacific Ocean. Because of the immenseness of the North Pacific and the extremely long residence time (more than 500 years) of the water, enormous quantities of North Pacific Deep Water can be produced by vertical mixing.

Considerable volumes of cold water generally of low salinity are formed in the Southern Ocean. Such water masses spread into the interior of the global ocean and to a large extent are responsible for the anomalous cold, low-salinity state of the modern oceans. The circumstances leading to this role for the Southern Ocean are related to the existence of a deep-ocean circumpolar belt around Antarctica that was established some 25 million years ago by the shifting lithospheric plates which make up Earth's surface. This belt establishes the Antarctic Circumpolar Current, which isolates Antarctica from the warm surface waters of the subtropics. The Antarctic Circumpolar Current does not completely sever contact with the lower latitudes. The Southern Ocean does have access to the waters of the north, but through deep- and bottom-water pathways. The basic dynamics of the Antarctic Circumpolar Current lift dense deep water occurring north of the current to the ocean surface south of it. Once exposed to the cold Antarctic air masses, the upwelling deep water is converted to the cold Antarctic Bottom Water and Antarctic Intermediate Water. The southward and upwelling deep water, which carries heat injected into the deep ocean by processes farther north, is balanced by the northward spread of cooler, fresher, oxygenated water masses of the Southern Ocean. It is estimated that the overturning rate of water south of the Antarctic Circumpolar Current amounts to 35 to 45 million cubic metres (1.2 to 1.6 billion cubic feet) per second, most of which becomes Antarctic Bottom Water.

The primary site of Antarctic Bottom Water formation is within the continental margins of the Weddell Sea, though some is produced in other coastal regions, such as the Ross Sea. Also, there is evidence of deep convective overturning farther offshore. Antarctic Bottom Water, formed at a rate of 30 million cubic metres per second, slips below the Antarctic Circumpolar Current and spreads to regions well north of the Equator. Slowly upwelling and modified by mixing with less dense water, it returns to the Southern Ocean as deep water.

The remaining upwelling of deep water spreads near the surface to the north, where it forms Antarctic Intermediate Water within the Antarctic Circumpolar Current zone and spreads along the base of the thermoclines farther north. This water mass forms a sheet of low-salinity water that demarcates the lower boundary of the subtropical thermocline. It upwells into the thermocline, partly compensating for the sinking of North Atlantic Deep Water.

Ocean Surface Waves

Ocean surface waves are surface waves that occur at the surface of an ocean.

They usually result from distant winds or geologic effects and may travel thousands of miles before striking land.

They range in size from small ripples to huge tsunamis.

There is surprisingly little actual forward motion of individual water particles in a wave, despite the large amount of forward energy it may carry.

The great majority of waves one sees on an ocean beach result from distant winds.

Three factors influence the formation of "wind waves": Windspeed; length of time the wind has blown over a given area; and distance of open water that the wind has blown over (called fetch).

References

- Pinet, Paul R. (1996). Invitation to Oceanography (3rd ed.). St. Paul, MN: West Publishing Co. ISBN 0-7637-2136-0

- Miller, Inglis J., Jr.; Mooser, Gregory (Jul 1979). "Taste Responses to Deuterium Oxide". Physiology & Behavior. 23 (1): 69–74. doi:10.1016/0031-9384(79)90124-0

- The-physical-properties-of-water, water-as-a-resource, water-knowledge: eniscuola.net, Retrieved 12 June 2018

- "Experimenter Drinks 'Heavy Water' at $5,000 a Quart". Popular Science Monthly. 126(4). New York: Popular Science Publishing. Apr 1935. p. 17. Retrieved 7 Jan 2011

- Braun, Charles L.; Smirnov, Sergei N. (1993), "Why is water blue?" (PDF), Journal of Chemical Education, 70 (8): 612–614, Bibcode:1993JChEd..70..612B, doi:10.1021/ed070p612

- Water-cycle: newworldencyclopedia.org, Retrieved 12 June 2018

- Hamblin, W. Kenneth; Christiansen, Eric H. (1998). Earth's Dynamic Systems (8th ed.). Upper Saddle River: Prentice-Hall. ISBN 0-13-018371-7

- Lewis, G. N.; MacDonald, R. T. (1933). "Concentration of H2 Isotope". The Journal of Chemical Physics. 1 (6): 341. Bibcode:1933JChPh...1..341L. doi:10.1063/1.1749300.

- Ocean-current, science: britannica.com, Retrieved 20 May 2018

- "Joseph Louis Gay-Lussac, French chemist (1778–1850)". 1902 Encyclopedia. Footnote 122-1. Retrieved 2016-05-26

- Murphy, D. M. (2005). "Review of the vapour pressures of ice and supercooled water for atmospheric applications". Quarterly Journal of the Royal Meteorological Society. 131: 1539–1565. Bibcode:2005QJRMS.131.1539M. doi:10.1256/qj.04.94.

Permissions

Index

A

Abiogenesis, 117-121, 149

Antarctic Circumpolar, 216, 220, 231, 233-234

Atmospheric Pressure, 41-42, 52, 93-95, 101-102, 189, 196, 229

Atmospheric Temperature, 18, 93, 102-104, 110

B

Biogeochemical Cycle, 116, 133, 140

Biogeography, 1-2, 30-34, 48, 62-63

C

Canyon, 6, 160, 181, 188

Carbon Cycle, 133-139, 147

Cartography, 45, 70, 88

Cave, 6, 25, 35

Climate Change, 20-21, 23, 30, 34, 40, 48-49, 93, 112-115, 128, 138, 222

Climatology, 1, 29, 31, 38, 40, 48-50, 62, 66

Condensation, 107, 196, 202, 205-206, 211

Continental Drift, 31, 166, 168, 176-178, 180

Contour Line, 74

Coriolis Effect, 57, 217, 226

D

Deforestation, 66, 112, 114, 208

Denitrification, 112, 141-143

Deposition, 5, 23, 59, 158-159, 162-163, 188, 190-191

E

Earthquake, 25, 179, 182-183

Ecological Pyramid, 116, 128, 131

Ekman Transport, 217, 228-231

Entrainment, 188-191

Environmental Geography, 1, 62

Erosion, 18, 23-24, 27-28, 37, 40, 122, 124, 148-149, 151, 158, 163-164, 186, 188-191, 202, 223

Extrusive Igneous Rock, 151

G

Geomorphology, 1, 23-24, 26, 29-30, 52, 62, 158, 192

Glaciology, 23, 51-52

Gps, 64, 80-82, 89-92, 175

Greenhouse Effect, 93, 103-104, 106-108, 110-111, 113

Greenhouse Gas, 21, 50, 111-112, 114, 135

Gulf Stream, 57, 220-221, 224-225, 230

H

Heat Capacity, 194-195, 199, 214, 218

Heat Flux, 218, 221

Hill, 7, 228

Hydrogen Bonding, 194, 198-199, 201

Hydrologic Cycle, 36-39, 202, 209-211

Hydrology, 2, 23-24, 29, 36-39, 44-45, 62, 85, 202

I

Intrusive Igneous Rock, 151

Ionosphere, 65, 91, 94, 101

Island Biogeography, 1, 32-33

M

Magnetic Striping, 177-178, 181

Magnetosphere, 94, 97, 101

Maritime, 45, 55, 218

Mesosphere, 94, 97, 100-103

Metamorphic Rock, 2, 145, 153, 164-166

Meteorology, 23, 29, 38, 40-47, 62, 66

Mountain, 7-8, 18, 25, 34, 53, 58, 60-62, 102, 151, 167, 170-172, 177, 184-185, 188, 192, 207

Mountain Building, 151, 172, 184, 188

Mudrock, 161

N

Nitrification, 141-143

Nitrogen Cycle, 133, 140-143

Nitrogen Fixation, 93, 141-143

O

Ocean Current, 193, 221, 223, 228

Oceanography, 23, 48, 54-57, 62-63, 193, 213-214, 234

Oparin-haldane Theory, 117-119

Ozone Layer, 93-94, 97, 99-100, 136

P

Palaeogeography, 23, 29, 58, 158

Paleoclimatology, 1, 29, 49, 56, 59-60, 158

Pedology, 28, 158

Photogrammetry, 64-65, 69, 82, 92

Phylogeny, 31-32

Phylogeography, 1, 33-36

Physical Oceanography, 55, 193, 213-214

Plate Tectonic, 58, 60, 168, 178

Plateau, 5, 7, 61, 172

Pressure Gradient, 226-227, 229

Pteridophyta, 11-13

R

Remote Sensing, 23, 40, 64-69, 82, 92

Renewable Energy, 46, 84

Rock Cycle, 144, 151-152, 165

S

Sandstone, 77, 145-146, 153, 159-160, 162-163, 166

Sedimentary Rock, 59, 145-147, 152-153, 158-159, 161, 163

Spatial Analysis, 77, 83, 85

Species Diversity, 15, 32-33, 121-123

Stratosphere, 94, 98-100, 102-103, 108, 144

Subsidence, 24-25, 171, 213

Subtropical Gyre, 230

Surface Tension, 193, 200-201

T

Thermohaline Circulation, 57, 196, 216, 219, 225, 229, 231-232

Thermosphere, 94, 101-103

Topographic Map, 73-76

Troposphere, 18, 40, 91, 94, 98-99, 102-103, 108, 219

V

Valley, 6-8, 53, 64, 75, 170-171, 180, 188

Vaporization, 107, 194-195

Vascular Plant, 13

Volcanic Eruption, 99, 124, 151, 153, 157-158, 184

Volcano, 8, 25, 44, 55, 124, 183-184

W

Water Cycle, 18, 97, 133, 193, 202-206, 208, 211

Weathering, 6, 24, 53, 123, 147, 151, 158, 160, 163, 186-190, 192

Wind-driven Circulation, 224-227, 229